Praise for *The Quick Guide to Prompt Engineering*

"It's not just a 'quick' guide; it's the *best* guide to accelerate your GenAI journey."

—Robert C. Wolcott,
Co-Founder & Chair, TWIN Global
Adjunct Professor of Innovation
University of Chicago & Northwestern University

"We need a fresh definition for computer literacy that is not only focused on the ability to use computers but also on how to ask computers the right questions that will ensure getting the best answers. Ian's new book is a practical guide that helps readers understand, learn, and implement prompt engineering successfully."

—Dina Fares,
Director of Digital Transformation
Digital Ajman

"Ian Khan has written a highly accessible book for understanding how to quickly use advanced skills to get the most value from popular generative AI solutions. The book can be used both as a reference guide to quickly look up specific techniques or even read from front to back to get a good handle on many of the most important and developing technologies and concepts in the AI field today. Ian's guide certainly has benefits for both beginners and advanced generative AI users."

—Dr. Jonathan Reichental,
Founder of Human Future,
Professor, and Author

"Prompting is an essential part of digital literacy or basic digital skills by now – if you have not mastered it yet, this book is a good start."

—Siim Sikkut,
Former CIO Government of Estonia.
Author of "Digital Government Excellence"

"Ian Khan has masterfully presented 'prompt engineering' as the key to human engagement with AI. It is truly an authoritative and comprehensive guide."

—Rafi-uddin Shikoh,
CEO, Dinar Standard

"Ian Khan has once again showcased his expertise and passion for helping others understand the technology that is transforming our lives. He adeptly elucidates the intricate subjects of artificial intelligence and prompt engineering in a manner that is not only understandable but, more importantly, practical. The Guide to Prompt Engineering stands as a crucial resource for comprehending the terminology, foundational elements, and the commercial and ethical considerations associated with the utilization of AI."

—Deborah Westphal,
Author of Convergence

The Quick Guide to

The Quick Guide to

Generative AI Tips and Tricks for
ChatGPT, Bard, Dall-E, and Midjourney

Ian Khan

Published by John Wiley & Sons, Inc., Hoboken, New Jersey.
Published simultaneously in Canada.

For general information on our other products and services or for technical support, please contact our Customer Care Department within the United States at (800) 762-2974, outside the United States at (317) 572-3993 or fax (317) 572-4002.

Wiley also publishes its books in a variety of electronic formats. Some content that appears in print may not be available in electronic formats. For more information about Wiley products, visit our web site at www.wiley.com.

Library of Congress Cataloging-in-Publication Data:

Names: Khan, Ian, author.
Title: The quick guide to prompt engineering / Ian Khan.
Description: First edition. | Hoboken, New Jersey : Wiley, [2024]
Identifiers: LCCN 2023055094 (print) | LCCN 2023055095 (ebook) | ISBN 9781394243327 (paperback) | ISBN 9781394243341 (adobe pdf) | ISBN 9781394243334 (epub)
Subjects: LCSH: Artificial intelligence. | Electronic data processing–Data preparation.
Classification: LCC Q336 .K43 2024 (print) | LCC Q336 (ebook) | DDC 006.3–dc23/eng/20240104
LC record available at https://lccn.loc.gov/2023055094
LC ebook record available at https://lccn.loc.gov/2023055095

Cover Design: Wiley
Cover Image: © siraanamwong/Adobe Stock Photos

SKY10066411_021724

To the pursuit of knowledge, innovation, and the adoption of new technologies.

Contents

Preface

WELCOME TO *The Quick Guide To Prompt Engineering*, a book I have passionately crafted to guide you through the transformative world of generative artificial intelligence (AI). As we stand on the brink of a technological revolution, this guide emerges as a crucial beacon of knowledge and insight.

Why This Book Now?

We are at a pivotal moment in the evolution of AI. The concepts explored in this book are at the forefront of this evolution, marking the beginning of what I believe to be the "Golden Era" of generative AI. In the next three to five years, we will witness unprecedented growth and innovation in this field, opening up a myriad of possibilities that will significantly shape our future.

In such a rapidly advancing landscape, acquiring new skills is not just beneficial; it's imperative. Understanding the workings of AI is no less than acquiring a superpower, one that will be invaluable for current and future generations. My aim with this book is to empower you with this superpower.

As we delve into this guide, you'll find that it's more than just a book; it's a journey into the future of technology. It's about understanding and harnessing the potential of AI to enhance human creativity and capability. Whether you are a student, a researcher, an industry professional, or simply an AI enthusiast, this book is your compass through the evolving landscape of generative AI and prompt engineering.

I invite you to join me on this exciting journey. Together, let's explore the myriad ways in which generative AI can transform our world, and let's equip ourselves with the knowledge and skills to be active participants in this transformation. In my humble opinion, this book is not just a recommendation; it's an essential read for everyone who wishes to be part of the AI-driven future.

—Ian Khan
January 8, 2024, New York

The Basics of Generative Artificial Intelligence

Table of Contents

Understanding AI, Machine Learning, and Deep Learning

What Is AI

Artificial intelligence (AI) is a branch of computer science that aims to create machines capable of mimicking human intelligence. Unlike traditional systems that follow explicit instructions, AI systems are designed to process information and make decisions or predictions based on the data they're given. The overarching goal of AI is to develop algorithms and models that allow machines to perform tasks—ranging from recognizing patterns to decision-making—that would usually require human cognition. AI's scope spans various technologies, including robotics, natural language processing (NLP), and expert systems. Its applications are evident in daily life, with systems such as virtual assistants, facial recognition software, and autonomous vehicles. AI's impact is transformative, redefining how industries operate and how we interact with technology.

Historical Development The journey of AI began in the 1940s and 1950s with the development of the first electronic computers. The 1980s saw the rise of machine learning (ML), where algorithms learn directly from data rather than relying on explicit programming. Neural networks, a subset of ML, faced challenges until the 2000s when computational power and data availability grew. This resurgence, now termed deep learning, uses multilayered neural networks to process vast datasets. The game-changing breakthroughs, such as Deep Blue's chess victory in 1997 and AlphaGo's win in 2016, marked significant milestones. Today, AI encompasses a blend of these techniques, continuously evolving with advancements in computation, data, and algorithms.

Applications of AI AI has woven its way into a multitude of sectors, revolutionizing processes and augmenting human capabilities. In health care, AI algorithms are being used to diagnose diseases, sometimes with accuracy surpassing human doctors. In finance, it powers fraud detection systems, optimizing security. The automotive industry is witnessing a transformation with AI-driven autonomous vehicles. In entertainment, recommendation systems such as those in Netflix or Spotify customize user experiences. E-commerce platforms use AI for predicting consumer behavior, enhancing sales strategies. Virtual assistants such as Siri and Alexa employ AI to comprehend and respond to user commands. In manufacturing, AI-driven robots optimize assembly lines, increasing efficiency. Additionally, in the realm of research, AI aids in complex simulations and data analysis. From smart homes to predictive text on smartphones, the applications of AI are vast, continuously expanding, and making an indelible mark on how society functions and evolves.

AI Today The current landscape of AI is characterized by rapid advancements and widespread adoption across various sectors. Breakthroughs in machine learning, especially deep learning, have propelled AI capabilities, making tasks such as image and speech recognition more accurate than ever before. AI models, such as GPT-3 and BERT, have revolutionized natural language processing, enabling seamless human-computer interactions. The growth of big data and enhanced computational power, through GPUs, has further accelerated AI research and applications. Today's businesses leverage AI for predictive analytics, customer insights, and automation. Ethical concerns, such as biases in AI models and privacy issues, have prompted discussions and regulations. Innovations in AI have also sparked debates on the future of employment, as automation replaces certain job functions. However, alongside challenges, AI offers immense potential to drive efficiency, innovation, and growth in the 21st century.

The Future of AI The future of AI holds immense potential and is poised to be transformational across various domains. As AI algorithms become more sophisticated, we'll see further personalization in services, from tailored education platforms to individualized health monitoring. The continued convergence of AI with fields such as quantum computing could redefine computational limits, allowing for the solving of currently insurmountable problems. Ethical considerations will gain prominence, with emphasis on transparency, fairness, and avoiding biases in AI systems. There will also be a focus on achieving general AI, a system with cognitive abilities akin to human intelligence. As AI integrates more deeply with our daily lives, new job roles and industries will emerge, while others adapt or phase out. Lastly, international collaborations and regulations will play a crucial role in ensuring AI's safe and equitable development and deployment.

What Is Machine Learning

Machine learning (ML) is a subset of artificial intelligence that focuses on the development of algorithms that allow computers to learn from and make decisions based on data. Rather than being explicitly programmed for a specific task, ML models use statistical techniques to understand patterns in data. By processing large amounts of data, these models can make predictions or decisions without human intervention. For example, a machine learning model can be trained to recognize images of cats by being shown many images of cats and non-cats. Over time, it fine-tunes its understanding and improves its accuracy. The essence of ML lies in its iterative nature; as more data becomes available, the model adjusts and evolves. This ability to learn from data makes machine learning integral in today's AI-driven world, fueling advancements in fields ranging from health care to finance.

What Is Deep Learning

Deep learning is a specialized subset of machine learning inspired by the structure and function of the human brain, specifically neural networks. It employs artificial neural networks, especially deep neural networks with multiple layers, to analyze various factors of data. Deep learning models are particularly powerful for tasks such as image and speech recognition. For instance, when processing an image, the model might first identify edges, then shapes, and eventually complex features such as faces or objects. The "deep" in deep learning refers to the number of layers in the neural network. Traditional neural networks might contain two or three layers, while deep networks can have hundreds. These intricate architectures allow deep learning models to automatically extract features and learn intricate patterns from vast amounts of data, often outperforming other machine learning models in accuracy and efficiency, especially when dealing with large-scale data.

What Is Generative AI

Generative AI refers to a subset of artificial intelligence models that are designed to generate new data samples that are similar in nature to a given set of input data. In essence, these models "learn" the underlying patterns, structures, and features of input data and then use this knowledge to create entirely new data samples. The resulting outputs, whether they are images, texts, or sounds, are often indistinguishable from real-world data. A quintessential example is the generative adversarial network (GAN), where two neural networks—a generator and a discriminator—are pitted against each other. The generator strives to produce data, while the discriminator evaluates its authenticity. Through iterative training, the generator improves its outputs. Beyond GANs, other generative models such as variational autoencoders (VAEs) also find extensive applications in tasks such as image synthesis and style transfer. The appeal of generative AI lies in its potential to craft novel yet coherent creations by understanding and mimicking complex data distributions.

Early Beginnings of Generative AI The genesis of generative AI dates back to the mid-20th century, rooted in foundational statistical modeling and pattern recognition techniques. Early forms of generative models included Gaussian mixture models (GMMs) and hidden Markov models (HMMs), which were pivotal in speech recognition and computational biology. While these models demonstrated the concept of capturing data distributions, their real-world applications were somewhat limited due to computational constraints and the lack of vast datasets. However, the introduction of neural networks in the 1980s paved the way for more sophisticated generative models. The Boltzmann machine, an early form of a neural network with a generative structure, was one such breakthrough. By the 2000s, with the rise of computational power and the availability of large datasets, models such as restricted Boltzmann machines (RBMs) became feasible. These foundational steps were the precursors to the contemporary generative models, such as GANs and VAEs, which now drive much of today's AI-generated content.

The Current Evolution of Generative AI Generative AI has experienced remarkable evolution in recent years, driven largely by advancements in neural network architectures and computational power. One of the pivotal moments was the introduction of GANs by Ian Goodfellow in 2014. As previously explained, GANs consist of two neural networks, the generator and discriminator, which work in tandem to produce highly realistic outputs. Variational autoencoders (VAEs) have also become a popular generative model, known for their probabilistic approach to generating new samples. These tools have facilitated groundbreaking applications such as creating realistic images, designing drug molecules, and even generating art and music. The surge in deepfake technology, which convincingly replaces faces in videos, underscores the power of these generative models. Additionally, transformer-based models, such as OpenAI's GPT series, have demonstrated the capability to generate humanlike text. The rapid progress in generative AI underscores its transformative potential and continuously blurs the line between human-generated and machine-generated content.

What Are Discriminative Models Discriminative models, in the realm of machine learning, are primarily concerned with distinguishing between different classes or categories based on input data. Rather than capturing the data distribution like generative models, they focus on modeling the boundary separating different classes. For instance, in a binary classification problem, a discriminative model would aim to discern the boundary that separates two categories, enabling predictions about which class a new input belongs to. Common examples of discriminative algorithms include logistic regression, support vector machines, and most deep neural networks designed for classification tasks. They are often chosen for tasks where pinpointing the exact decision boundary is more crucial than understanding the underlying data distribution. Discriminative models, given their direct approach, tend to be more accurate than generative models for classification tasks, but they don't offer insights into the characteristics or patterns that define each class.

Applications of Generative AI Generative AI has revolutionized numerous fields with its ability to generate new, previously unseen content. In art and entertainment, GANs have been utilized to create realistic artwork, music, and even video game levels. In the fashion industry, generative models suggest novel clothing designs or adapt existing styles to personalized preferences. The health care sector benefits from synthesizing medical images for research, enhancing the training data pool without compromising patient privacy. In the realm of natural language processing, generative models, such as GPT variants, produce humanlike text, enabling more sophisticated chatbots and content creation tools. Additionally, in the realm of chemistry and drug discovery, generative models propose molecular structures for new potential drugs. Generative AI also aids in data augmentation, where limited datasets are expanded by creating variations, thus improving model training. These applications underscore generative AI's transformative potential across diverse sectors.

Limitations of Generative AI Generative AI, despite its groundbreaking capabilities, possesses inherent limitations. Firstly, training generative models, especially advanced architectures such as GANs, demands considerable computational resources and time. This is not always feasible for individual developers or small entities. Secondly, these models can sometimes produce unrealistic or nonsensical outputs, especially when they encounter data significantly different from their training set. Another concern is the ethical implications of generative

AI: the creation of deepfakes in videos or misleading information can have severe societal ramifications. Intellectual property rights can also be jeopardized when generative models produce content indistinguishable from human-made creations. Moreover, ensuring fairness and avoiding biases in outputs is challenging, as these models can inadvertently learn and perpetuate existing biases from their training data. Lastly, interpretability remains a challenge; understanding how these models arrive at particular outputs is not always straightforward, which can hinder trust and widespread adoption.

The Future of Generative AI Generative AI stands at the precipice of a transformative future, redefining various industries and societal interactions. As computational power advances and algorithms refine, we anticipate more robust and efficient generative models. These models will likely produce outputs of higher fidelity, increasing their realism and utility. Integration with augmented reality (AR) and virtual reality (VR) environments could revolutionize the entertainment, gaming, and education sectors. Custom content creation, tailored to individual preferences, will become commonplace, personalizing user experiences like never before. Ethical considerations will take center stage, prompting the development of regulatory frameworks and tools to detect AI-generated content, combating misinformation and unauthorized reproductions. Additionally, advancements in semi-supervised and unsupervised learning will make generative AI more accessible, reducing the need for vast labeled datasets. Collaborative efforts between AI researchers and domain experts will further broaden the horizons, unlocking multifaceted applications that are currently unforeseen.

What Is a Language Model?

Language models have undergone significant advancements over the past few years. At their core, these models are designed to understand and generate human language. Through different architectural approaches and training methods, researchers have developed several types of language models, each catering to specific needs and applications.

N-gram Language Models This is one of the earliest types of language models. An n-gram model predicts the next word in a sequence based on the (n – 1) preceding words. For instance, a bigram (2-gram) model would consider two words at a time.

Usage: N-gram models have been historically used in spell-check systems and basic text predictions.

Limitation: These models struggle with long-term dependencies because they only consider the *n* previous words. Additionally, they do not scale well with increasing vocabulary sizes.

Recurrent Neural Networks (RNNs) RNNs process sequences of data by maintaining a *memory* from previous steps. This allows them to capture information from earlier in the sequence and use it to influence later predictions.

Usage: RNNs have been employed in tasks such as machine translation, and sentiment analysis.

Limitation: They can be computationally intensive and face challenges with very long sequences, often forgetting information from the earliest parts of the input.

Long Short-Term Memory (LSTM) Networks LSTM is a special kind of RNN that includes a mechanism to remember and forget information selectively. This helps in tackling the long-term dependency problem seen in basic RNNs.

Usage: LSTMs are widely used in time series forecasting, machine translation, and speech recognition.

Limitation: While LSTMs mitigate some of the challenges of RNNs, they can still be computationally heavy, especially with very large datasets.

Transformer Models Introduced in the paper "Attention Is All You Need," transformer models utilize self-attention mechanisms to weigh input data differently, enabling the model to focus on more relevant parts of the input for different tasks.

Usage: Transformers have become the go-to architecture for many NLP tasks, including text generation, machine translation, and question answering.

Limitation: The computational needs for transformer models are intense, necessitating powerful hardware setups, especially for large-scale models.

BERT (Bidirectional Encoder Representations from Transformers) BERT is a pretrained transformer model that considers the context from both the left and the right side of a word in all layers, making it deeply bidirectional.

Usage: BERT and its variants have set state-of-the-art performance records on several NLP tasks such as sentiment analysis and named entity recognition.

Limitation: Fine-tuning BERT for specific tasks can be computationally expensive. Additionally, its deep bidirectionality can make it less interpretable.

GPT (Generative Pretrained Transformer) Unlike BERT, which is trained to predict masked words in a sequence, GPT is trained to predict the next word in a sequence, making it a generative model.

Usage: GPT models, especially GPT-3 by OpenAI, have demonstrated humanlike text generation capabilities, answering questions, writing essays, and even crafting poetry.

Limitation: GPT models can sometimes generate plausible-sounding but incorrect or nonsensical outputs. They also require vast amounts of data for training.

Summary Language models have transitioned from simple statistical methods to complex neural network architectures. With each evolution, they've become more adept at understanding the intricacies of human language. However, each model type has its strengths and challenges, and the choice often depends on the specific application and available computational resources. As AI research advances, we can anticipate even more sophisticated models that seamlessly integrate with human linguistic interactions.

Applications in Data Management Data management, the practice of collecting, keeping, and using data securely, efficiently, and cost-effectively, is essential to businesses and organizations of all sizes. With the recent rise of sophisticated language models, there's been a transformative shift in how data management processes are executed. Here's a look at how language models are revolutionizing data management:

Data Entry and Cleaning Manual data entry and data cleaning are two of the most time-consuming tasks in data management. Language models can automate these processes by extracting information from unstructured sources such as emails, documents, and websites, converting them into structured formats. Additionally, they can identify and rectify inconsistencies, duplicates, and errors in datasets, ensuring data quality.

Semantic Search Traditional search mechanisms rely on keyword matching, often returning irrelevant results. With language models, semantic search becomes possible, wherein the context and meaning of the query are understood. This ensures that database searches are not just keyword-based but contextually relevant, fetching more accurate and meaningful results.

Data Classification and Categorization Language models can automatically categorize and label vast amounts of data. For instance, customer feedback can be automatically sorted into categories such as positive, negative, or neutral. Similarly, documents can be classified based on their content, facilitating faster retrieval and better organization.

Natural Language Queries For those unfamiliar with SQL or other database querying languages, extracting specific data can be challenging. Language models allow users to fetch data using natural language queries. For instance, a user could ask, "Show me sales data for the last quarter," and the language model would translate that into an appropriate database query.

Content Generation and Summarization Language models can generate humanlike text based on data insights. For businesses, this could mean automatic report generation, where insights drawn from data analytics are converted into understandable narratives. Additionally, models can summarize vast amounts of data, providing executives with concise briefs instead of lengthy reports.

Data Privacy and Redaction With rising concerns about data privacy, there's an increasing need to redact personal information from databases, especially when sharing datasets. Language models can automatically identify and mask sensitive information, ensuring data privacy compliance.

Chatbots and Customer Support Data management isn't just about handling internal data but also managing customer interactions. Language models power intelligent chatbots that can fetch information from databases in real time to answer customer queries, reducing the load on human agents and ensuring efficient data-driven customer service.

Predictive Text and Autocompletion For data managers and analysts, predictive text powered by language models can expedite data entry tasks. By predicting what the user intends to type next, these models can accelerate the data entry process, reducing manual effort and errors.

Multilingual Data Management In a globalized world, businesses often deal with data in multiple languages. Language models can automatically translate and transcribe data, ensuring seamless data management across linguistic barriers.

Insights and Recommendations Language models, when combined with other AI techniques, can provide actionable insights by analyzing patterns and trends in data. For e-commerce businesses, this could mean product recommendations based on customer behavior and preferences.

In conclusion, language models are rapidly becoming a cornerstone of modern data management. By automating tasks, ensuring data quality, and facilitating human-AI collaboration, these models are streamlining data processes and enabling businesses to derive more value from their data. As they continue to evolve, the synergy between language models and data management promises even more innovative solutions and efficiencies.

Applications of AI in Business

Health Care AI emerges as a transformative force in health care, offering unprecedented opportunities for both care delivery and business processes. Several ways AI has been instrumental in health care include:

Disease identification and diagnosis: Advanced AI algorithms analyze medical imaging such as X-rays, MRIs, and CT scans, aiding in the early detection and diagnosis of diseases such as cancer, allowing for timely interventions.

Treatment personalization: AI analyzes patient data to recommend personalized treatment plans, taking into account the patient's genetic makeup, lifestyle, and other factors.

Drug discovery and development: AI accelerates the drug development process by predicting how different compounds can treat diseases, significantly reducing the time and cost associated with traditional research.

Operational efficiency: AI-powered systems streamline administrative tasks such as appointment scheduling, billing, and patient record maintenance, leading to enhanced operational efficiency.

Remote monitoring: Wearable devices equipped with AI monitor vital statistics, alerting health care providers to potential health issues, enabling early intervention and reducing hospital readmissions.

For businesses within the health care sector, embracing AI equates to improved patient outcomes, reduced costs, and optimized operations. As AI continues to evolve, its potential to reshape health care delivery and its associated business models becomes increasingly evident.

Manufacturing AI stands at the forefront of the Fourth Industrial Revolution, reshaping the manufacturing landscape. The integration of AI in manufacturing yields several transformative benefits:

Predictive maintenance: AI systems analyze machine data to predict when equipment is likely to fail, enabling timely maintenance. This reduces downtime, extending machinery life and decreasing operational costs.

Quality assurance: Advanced vision systems powered by AI ensure product quality by identifying defects in real time on the production line, guaranteeing consistent product quality and reducing wastage.

Supply chain optimization: AI algorithms process vast amounts of data to optimize inventory levels, predict demand, and enhance supply chain agility.

Smart robotics: Robots, augmented with AI, can perform complex tasks, adapt to changes, and work collaboratively with humans, boosting production efficiency.

Energy consumption reduction: AI-driven systems monitor and analyze energy usage patterns, optimizing consumption and leading to significant cost savings.

For businesses in the manufacturing domain, AI represents an avenue for innovation, operational excellence, and cost-efficiency. Its continued integration is set to further elevate manufacturing capabilities, driving industry growth.

Disaster Management In the face of increasing global calamities, businesses are leveraging AI to fortify disaster management efforts, ensuring continuity, and safeguarding assets and human resources:

Early warning systems: AI models process vast amounts of data from satellites, ocean buoys, and sensors to predict natural disasters such as hurricanes, earthquakes, or floods, allowing businesses to implement precautionary measures in a timely manner.

Resource allocation: After a disaster, AI algorithms analyze the impact and distribute resources efficiently, ensuring urgent supplies reach the hardest-hit areas promptly.

Damage assessment: AI-driven drones and satellite imagery help in assessing the extent of damage, assisting businesses in understanding the immediate implications on infrastructure, operations, and supply chains.

Rescue operations: AI-enhanced robots are deployed in situations too hazardous for humans, ensuring swift rescue missions, especially in collapsed buildings or flood situations.

Business continuity planning: AI assists businesses in creating robust continuity plans by simulating disaster scenarios, ensuring minimal disruptions during real-world events.

For businesses, AI's application in disaster management isn't merely a technological advancement; it's a crucial strategy to ensure resilience, safety, and sustainability in a volatile world.

Climate Change Climate change presents a complex challenge, and businesses are turning to AI to both mitigate its effects and adapt to its evolving realities:

- **Predictive analysis:** Businesses are using AI to forecast environmental shifts and the implications they hold for industries. This helps firms in sectors such as agriculture, real estate, and insurance anticipate, prepare for, and navigate changes.
- **Carbon footprint reduction:** AI optimizes energy use in manufacturing processes, warehouses, and offices. By monitoring and

adjusting energy consumption patterns, companies can reduce emissions and operational costs.

- **Supply chain resilience:** AI algorithms predict climate-induced disruptions and suggest alternatives, ensuring businesses maintain seamless operations even under unpredictable weather patterns.
- **Sustainable solutions development:** AI is aiding research in sustainable materials and renewable energy. Companies in the energy sector use it to optimize the output of solar panels and wind turbines.
- **Stakeholder engagement:** Businesses employ AI to analyze consumer sentiment, enabling them to align products and marketing strategies with growing demand for sustainability.

In the fight against climate change, AI empowers businesses to be proactive, making them part of the solution while ensuring long-term sustainability and resilience.

Economy AI is shaping the economic landscape, redefining the way businesses operate and driving economic growth:

- **Efficiency and automation:** Businesses are adopting AI-driven automation to streamline operations, reduce overhead costs, and enhance productivity. This leads to optimized business processes and increased competitiveness in the global market.
- **Financial analysis:** AI algorithms provide deeper insights into market trends, predicting stock market movements, and assisting businesses in making informed investment decisions. Furthermore, fintech companies leverage AI for fraud detection and credit risk assessment.
- **Supply chain optimization:** AI assists businesses in predicting demand, ensuring optimal stock levels, and minimizing wastage. This results in a more agile and responsive supply chain, adapting to market shifts.

- **Consumer personalization:** AI-driven analytics enable businesses to understand consumer preferences in real time, allowing for personalized product recommendations, which boost sales and enhance customer loyalty.
- **Job creation and evolution:** While there's concern over AI displacing jobs, it's also creating new roles and reshaping existing ones. Businesses are benefiting from a skilled workforce trained to harness the capabilities of AI.

In summary, AI acts as a catalyst in the economic sphere, promoting growth, enhancing efficiency, and redefining business operations.

The Role of Prompts in Generative AI

Table of Contents

How Did Prompts Originate

Prompts, in the simplest terms, are the initial stimuli or cues provided to a system to elicit a particular response. The idea of using prompts is not novel and dates back to the early days of computing and AI.

The genesis of prompts can be traced back to rule-based systems, where specific inputs led to predefined outputs. These were systems that operated on strict logic and deterministic patterns. If you asked them a question or gave them an instruction, they responded precisely in the way they were programmed to.

As we moved into the era of machine learning, datasets became the new prompts. Algorithms were trained on labeled datasets, where the data acted as a "prompt" to determine the appropriate label. Supervised learning, a dominant paradigm, essentially relied on feeding the system a series of prompts (input data) and desired outputs (labels). Over time, the model learned the patterns and was able to predict the output for new, unseen inputs.

With the advent of deep learning and, more specifically, generative models such as GANs and VAEs (variational autoencoders), the idea of prompting underwent a transformation. Here, one network often generates content while another evaluates it. In GANs, for instance, the generator is "prompted" by random noise to produce images, which are then evaluated by the discriminator.

The modern notion of prompts, especially in the context of language models such as OpenAI's GPT series, stems from the capability of these models to generate coherent, diverse, and contextually relevant outputs based on a given input string or "prompt." They are no longer just a command but a creative nudge that guides the vast potential of the model in a particular direction.

The ubiquity of prompts in generative AI today is a culmination of decades of evolution in computing paradigms. From rigid rule-based systems to the flexible and creative AI models of today, prompts have consistently played a crucial role in shaping system outputs. Their origin story underscores the ever-present human desire to communicate with, guide, and derive utility from machines in meaningful ways.

How Can You Provide Data Input to an AI System

Data input is the backbone of any AI system. The type, quality, and format of the data you input can significantly influence the performance and output of the system. Given the vast landscape of AI, the methodologies for data input can vary based on the specific application, but some universal principles and methods apply.

Manual entry: At its most basic, data can be inputted manually into a system. This method is common for systems with simple user interfaces, such as chatbots or search engines. Users type queries or commands, and the AI responds accordingly.

Structured databases: For more complex tasks, such as in business analytics or customer relationship management, AI systems often draw data from structured databases. These databases, often relational, store data in tables, making it easy for AI algorithms to query and process.

Data streams: In real-time applications such as stock trading or traffic management, AI systems tap into continuous data streams. This streaming data can come from various sources, including sensors, cameras, and online feeds.

APIs (application programming interfaces): APIs allow different software systems to communicate. AI systems can pull data from other platforms or services via APIs, ensuring dynamic data exchange and up-to-date information. For instance, language models might access current weather data through a weather API.

File uploads: In scenarios such as image recognition or document analysis, users can provide data by uploading specific files. These could be image files, PDFs, audio files, or any other format relevant to the task at hand.

Web scraping: Some AI projects, especially those requiring vast amounts of data from the Internet (such as sentiment analysis), employ web scraping tools. These tools automatically extract data from web pages, feeding it to the AI system for analysis.

Interactive prompts: Especially prevalent in generative models, users can give a prompt or seed input. For instance, when working with text-generating models, a user might input a sentence or phrase, and the AI will continue or elaborate based on its training.

Sensors and IoT devices: The Internet of Things (IoT) has enabled AI systems to receive data directly from the physical world. This data can come from wearable devices, home automation systems, industrial machinery, and more. It's especially crucial for applications in health monitoring or smart cities.

In conclusion, the method of data input largely depends on the specific requirements and nature of the AI application. While some methods are passive, where AI continuously receives data, others are more active, requiring user intervention. Regardless of the method, it's crucial to ensure that the data is relevant, clean, and unbiased to make the most of the AI system's capabilities.

Making AI Accessible to Everyone

Generative AI has ushered in an era of unparalleled innovation, but its true strength lies not just in its technical prowess but in democratizing access. The ability for diverse populations to leverage, understand, and benefit from AI has become a central discourse in the tech industry.

Here's how the role of prompts in generative AI contributes to this democratization

Simplicity and intuitiveness: The very nature of prompts is based on human language, making it accessible even to those without a technical background. Instead of mastering a programming language or complex interfaces, users can engage with AI models using simple text instructions.

Cost-effective interaction: With the traditional approach, utilizing AI often required costly hardware or specialized software. By relying on cloud-based generative models that use prompts, users can access powerful AI without significant investment, making it financially accessible.

Personalized outputs: Generative AI, through prompts, can be tailored to produce results that resonate with specific cultures, languages, or individual preferences. This flexibility ensures that AI isn't just a one-size-fits-all solution but can be molded to serve diverse populations.

Educational opportunities: Prompts offer an excellent avenue for educators to introduce students to the world of AI. Given its simplicity, students from various age groups can experiment, understand, and appreciate the capabilities and ethics surrounding AI.

Support for non-English speakers: Many generative models trained on diverse datasets understand multiple languages. This multilingual support ensures that non-English speakers can engage with and benefit from AI just as effectively.

Empowerment for entrepreneurs and SMEs: Small and medium-sized enterprises (SMEs) often lack the resources for extensive AI deployments. With prompt-based models, they can access top-tier AI capabilities to improve their operations, products, or services without the need for large teams or budgets.

Enhancing creativity: Artists, writers, and other creative professionals can use prompts to brainstorm, draft, or refine their work, ensuring that AI becomes a tool for augmenting human creativity rather than replacing it.

Community development: Open platforms that utilize generative models with prompts allow for community input, feedback, and development. This collective contribution ensures that the AI systems evolve in a direction that serves the broader population's interests and needs.

In conclusion, the introduction of prompts in generative AI is not just a technical advancement but a societal one. It bridges the gap between sophisticated technology and everyday users, ensuring that the benefits of AI are reaped by everyone, from industry professionals to students, from large corporations to individual artists. In this light, prompts aren't just a method of communication with AI; they're a step toward a more inclusive digital future.

How Prompts Guide the AI's Response

What Is behind the Prompt

Behind every prompt fed to an AI system lies a labyrinth of complexities, a confluence of algorithms, historical data, neural pathways, and contextual interpretations. Understanding these intricate mechanisms provides insight into how AI generates responses and navigates the vast universe of information.

One of the primary factors guiding AI's response to a prompt is its training data. The AI "remembers" a vast array of patterns, structures, and contextual information based on millions or even billions of data points it has been trained on. Each prompt is compared to this historical context to generate the most relevant and accurate answer.

Neural Architectures At the heart of AI's processing capabilities are neural networks—architectures inspired by human brain pathways. These networks, comprising layers of interconnected nodes, process prompts in stages. Each layer extracts and interprets different levels of information from the prompt, progressively refining the AI's understanding and subsequent response.

Tokenization and Vectorization Before AI can process a prompt, it must translate it into a language it understands. Prompts are broken down into tokens (often words or sub-words) and then converted into numerical vectors. This translation facilitates the AI's ability to discern relationships, context, and meaning from human language input.

Attention Mechanisms These are pivotal in helping AI systems focus on the most critical parts of a prompt. By assigning weights to various segments of the input, AI can prioritize and generate responses that emphasize the most relevant parts, effectively fine-tuning its answers to align closely with the user's intent.

Generative Capability After processing, the AI must reconstruct a coherent and contextually appropriate response. This generation process involves not just reproducing known patterns but also creatively assembling them in ways that make sense for the specific prompt at hand.

Adaptive Learning While deep learning models do not traditionally "learn" from each new prompt in real time, feedback loops in some systems allow for continual improvement. The AI system can be refined over time, adapting its response mechanisms based on new data or feedback.

Bias and Ethical Considerations Implicit in any prompt-response mechanism are the biases ingrained from training data. These biases can inadvertently shape the AI's response, making it crucial for developers and users to be aware of and mitigate potential skewed perspectives.

Behind the simple interface of providing a prompt and receiving an AI-generated response lies a dense web of processes and decisions. This intricate ballet of computation, combined with ever-evolving techniques, ensures AI not only comprehends our queries but also crafts answers that are both relevant and enlightening. As the field advances, so too will the depth and breadth of understanding behind each prompt.

How Do Generative AI Systems Understand Input and Provide Output

Generative AI systems have rapidly transformed our technological landscape, with their remarkable capability to understand complex inputs and generate diverse outputs. Delving into their mechanics reveals a fascinating dance of algorithms, data patterns, and intricate computations.

When a user feeds an input or prompt to a generative AI, the system begins by breaking it down into comprehensible units, often tokens, which can be words or sub-words. This tokenization ensures the system can analyze and process each fragment of information efficiently.

Here, each word or sub-word gets a numeric representation, capturing its semantic essence and relationship to other words.

Modern generative AI models, particularly transformers such as GPT-3 or BERT, use attention mechanisms. These mechanisms allow the model to weigh different parts of the input differently, focusing on the most crucial segments while considering the broader context. Essentially, the AI system determines which parts of the input are most relevant to generating an appropriate response.

At the heart of generative models lies a deep neural network. As the input travels through this network, each layer refines and redefines its understanding, using patterns learned during training. The depth of these layers facilitates the capture of complex structures and relationships within the input.

Once the input is comprehensively processed, the AI begins generating output, one piece at a time. This is often achieved through a probability distribution, where the AI system chooses the next word or token based on its maximum likelihood, given the current context.

Some advanced generative systems employ feedback loops, allowing real-time adaptation. By considering the generated output's context, the AI can refine subsequent portions of its response, ensuring coherence and relevance.

After creating an initial draft of the output, some models undergo postprocessing stages to refine and polish the generated content, ensuring it meets specific criteria or constraints.

Generative AI systems merge an intricate understanding of language, context, and data patterns to transform inputs into meaningful outputs. Their capability to comprehend, process, and produce content mirrors, to some extent, the cognitive processes in humans but at a computational scale and speed that has unlocked myriad applications in various fields.

What Goes behind the Scenes in a Generative AI System

A generative AI system's incredible prowess in producing content tailored to specific prompts is the culmination of sophisticated processes and computations. Behind the scenes, this journey from input to output is both intricate and enlightening.

Training on Massive Datasets Before a generative model can respond to prompts, it undergoes extensive training on vast datasets. This foundational step allows the model to learn patterns, structures, and nuances in language, granting it the capability to generate coherent and contextually relevant content.

Tokenization When a prompt is fed into the system, it's initially tokenized, breaking the content into manageable units (often words or sub-words). This process allows the AI to individually assess and process each segment.

Embedding and Vector Representation Tokens are then mapped into a high-dimensional space through embeddings. Each token is translated into a numerical vector, encapsulating its contextual and semantic essence in relation to other tokens.

Processing via Neural Networks The core of generative AI lies in its neural network, typically deep learning models such as transformers. These models contain millions, if not billions, of parameters. As token vectors traverse through the network's layers, complex operations identify relationships, patterns, and structures, refining the AI's understanding with each layer.

Attention Mechanisms Modern models employ attention mechanisms that enable them to weigh the significance of different parts of the input. By focusing on relevant segments and understanding broader context, AI systems can produce responses that are coherent and contextually apt.

Output Generation Leveraging learned patterns and the provided prompt, the AI produces output sequentially. It predicts the next token based on probability distributions, ensuring each added token aligns well with the existing content.

Decoding Strategies Generative models employ decoding strategies such as beam search or nucleus sampling. These strategies influence the diversity and quality of generated content, balancing between exploration (generating diverse content) and exploitation (sticking to more probable outputs).

Regularization and Optimization To prevent overfitting and ensure the model generalizes well to new prompts, regularization techniques are applied. Optimization algorithms adjust the model's parameters to minimize discrepancies between generated outputs and actual training data.

Behind every AI-generated response is a cascade of processes and computations, a testament to the power of modern machine learning. These systems, while automated, rely heavily on the vast amount of data they've been trained on and the intricate dance of algorithms that process, assess, and generate content.

The Importance of Carefully Engineering Prompts

The Need to Prepare an AI System with Information and Data The promise of AI is often tempered by a practical reality: these systems are only as insightful, accurate, and effective as the information they're provided. The adage "garbage in, garbage out," is particularly resonant in the realm of AI. Preparing an AI system with the right information and data is paramount for several pivotal reasons:

Ensuring Accuracy AI systems make decisions based on patterns in data. Feeding an AI model with comprehensive, accurate, and well-curated data ensures it makes informed and accurate decisions or predictions. This is especially crucial for applications such as medical diagnostics, financial forecasting, and autonomous vehicles, where inaccuracies can have life-altering consequences.

Mitigating Biases AI systems can inadvertently perpetuate or even amplify societal biases if trained on skewed or biased data. By carefully curating and preparing datasets and by being conscious of potential pitfalls, we can work toward models that are fairer and more equitable.

Improving Generalization An AI trained on diverse and extensive data can generalize better to unseen scenarios. This means it can handle a wider range of inputs and situations in real-world applications, thereby being more versatile and reliable.

Efficiency in Learning Properly prepared data can speed up the training process. Clean, balanced, and structured data can reduce the computational resources required and lead to faster convergence during model training.

Enhancing Model Interpretability When AI systems are primed with well-organized data, their predictions and actions become more interpretable. This is crucial for domains where understanding the why behind an AI's decision is as important as the decision itself, such as in legal or medical contexts.

Cost and Time Savings Preparing AI with the right information from the outset can reduce iterative adjustments later on. This leads to time and cost savings in the longer run, as less post-deployment tweaking is required.

User Trust and Adoption For users to trust and adopt AI solutions, they need to believe in the system's competency. Properly prepared AI models, informed by robust and relevant data, are more likely to win user trust and find widespread adoption.

In conclusion, the preparation of an AI system with comprehensive information and data isn't merely a technical requirement; it's an ethical and practical imperative. As AI's footprint expands across sectors and domains, the care with which we feed these systems will shape their efficacy, fairness, and societal impact.

The Need to Create Prompts to Receive the Best Output Crafting the right prompts for an AI system is akin to refining a question before seeking an answer. The quality and clarity of the prompt significantly influence the nature of the AI's response. Here's why careful engineering of prompts is vital:

Precision in Responses A well-engineered prompt narrows down the possible interpretations by the AI, ensuring a more precise and relevant response. Consider the difference between asking an AI, "Tell me about apples" versus "Describe the nutritional benefits of eating apples." The latter yields a more specific and focused answer.

Mitigating Misunderstandings Vague or ambiguous prompts can lead to AI outputs that may be irrelevant or even misleading. By being clear and specific in our prompts, we reduce the chances of misunderstanding and improve the reliability of the AI's output.

Efficient Interaction For users who seek swift and relevant answers, spending time sifting through off-target or too broad responses can be counterproductive. Well-engineered prompts save user time by getting to the heart of the matter quickly.

Reduced Risk of Biased Outputs The phrasing and content of prompts can inadvertently lead the AI toward biased or sensitive responses.

Thoughtfully constructed prompts can guide AI away from potential pitfalls and controversial outputs.

Facilitating Learning and Adaptation In systems where AI learns from interactions, well-framed prompts become even more crucial. They serve as clear signals to the model, aiding in more effective learning and adaptation over time.

User Trust and Satisfaction A user's trust in an AI system grows when they consistently receive accurate and relevant responses. Thoughtfully engineered prompts are instrumental in ensuring this consistency, fostering a sense of reliability and satisfaction among users.

Conserving Resources For businesses and developers, each interaction with an AI system may involve computational costs. Ensuring that prompts are crafted to extract the right response in the first instance can lead to savings in computational resources and costs.

In essence, the art of prompt engineering is central to harnessing the full potential of AI. It's not just about inputting data but doing so in a way that aligns with the desired outcome. As AI systems become more intricate and their applications more widespread, the nuanced craft of prompt creation will undoubtedly play an increasingly pivotal role in shaping effective AI-human interactions.

How Do Carefully Engineered Prompts Create a Good Output The efficacy of a generative AI's output is intimately tied to the input it receives. Carefully engineered prompts play a pivotal role in directing the AI's response in a way that is relevant, accurate, and contextually appropriate. Here's how meticulous prompt engineering facilitates superior outputs.

AI systems, especially language models, navigate vast amounts of information. A well-crafted prompt serves as a compass, guiding the AI to traverse its knowledge base in a specific direction, ensuring that the output aligns closely with the user's intention. Ambiguous prompts can lead the AI to produce generalized or irrelevant responses. By refining the prompt to be explicit and clear, we eliminate potential ambiguities, ensuring that the AI's output is on point and accurate.

A well-engineered prompt can also define the structure or format of the desired answer. For instance, specifying, "List the top three . . ." can ensure a concise, list-based response, while "Explain the process of . . ." may elicit a more detailed and explanatory answer. Thoughtfully constructed prompts can act as safeguards against unintentional biases. By being precise and neutral in phrasing, we can guide the AI to produce responses that are unbiased and factual.

A carefully designed prompt can provide context, enabling the AI to generate answers that are not just factually correct but also contextually appropriate. For instance, a prompt such as "Explain photosynthesis to a fifth-grader" ensures that the output is both accurate and accessible to the intended audience. Over time, observing how AI responds to different prompts can provide insights into its behavior. This iterative process allows users and developers to refine their prompt strategies, progressively improving the quality of the AI's output.

Carefully engineered prompts lead to more relevant and tailored responses, enhancing the user experience. When users receive accurate and contextually appropriate answers, their trust in the system grows, leading to increased engagement and reliance.

In conclusion, the relationship between the quality of a prompt and the AI's output is inextricable. In the realm of generative AI, the adage "garbage in, garbage out" holds true; the quality of input directly influences the quality of the output. As such, investing time and effort in crafting precise, clear, and thoughtful prompts is crucial for harnessing the full potential of AI systems and ensuring they serve their intended purpose effectively.

A Step-by-Step Guide to Creating Effective Prompts

Table of Contents

Prompts serve as a communication bridge between humans and AI systems. Crafting an effective prompt is essential to ensure that AI systems understand the user's intention and provide meaningful responses. Here are some common ways to create a prompt. In this chapter we will look at the very essentials of prompt generation or in other words providing commands to generative AI that are effective.

The first method is known as descriptive phrasing: Begin with clear and descriptive language. Instead of simply stating a keyword or topic, elaborate briefly. For instance, rather than prompting "Elephants," you could ask, "Provide information about the habitat and behavior of African elephants."

Use a question-based approach: Posing your prompt as a direct question can be an effective way to seek specific answers. For example, "What are the primary causes of climate change?" would likely yield a more focused response than a more ambiguous statement.

Think in terms of scenario framing: Set up a hypothetical scenario or situation. This is particularly useful when looking for detailed or process-oriented responses. An example could be, "Imagine you're teaching a high school class. Explain the process of photosynthesis in simple terms."

Ask for comparison requests: Asking the AI to compare two or more items can clarify complex topics. For instance, "Compare the economic impacts of solar energy vs. coal."

Use explicit formatting: If you're seeking a specific format in the answer, state it in the prompt. For example, "List down the top five most populous countries in the world."

The method you choose largely depends on your objective and the nature of the information you seek. The key is to be clear, specific, and provide enough context to guide the AI system. As AI continues to evolve, mastering the art of prompting becomes even more crucial in harnessing its full potential.

Various Platforms and Their Prompt Formats

Different generative AI platforms and tools often have unique formats or conventions for their prompts, tailored to their specific functionalities and objectives. Here's a look at some platforms and their respective prompt styles:

OpenAI's GPT models: OpenAI's models, especially those such as GPT-3, work well with natural language prompts. The prompts can be questions, statements, or scenarios. For instance, "Translate the following English text to French: 'Hello, how are you?'"

BERT and transformers: These models, used for tasks such as text classification or question answering, require prompts to be tokenized and passed as input sequences. For example, given a question and a passage, the prompt might be a combined token sequence of both.

Image classification (e.g. ResNet, VGG): Here, the "prompt" is typically an image, processed and standardized to fit the model's input dimensions. The user doesn't provide a text prompt but selects an image to be classified.

StyleGAN and image generators: In these models, the prompt could be a text description such as "a two-storied brick house in daylight," which the model tries to visually generate.

Music generators (e.g. MuseNet): Prompts might include a few notes, a genre, or a description, guiding the model to produce a specific style or continuation of music.

Tacotron for speech synthesis: The prompt is textual, e.g. "Read this text in a calm manner." The model then converts the text into speech, considering the provided instruction.

When crafting prompts for different platforms, it's essential to understand the platform's requirements and the underlying model's nature. Familiarizing oneself with the platform's documentation and guidelines is crucial to create effective prompts that produce desired outcomes. As generative AI systems become more complex, it is very much possible they will be able to understand easier prompts.

Recognizing Characters and Depth of Prompts

Crafting prompts for AI isn't just about the words used; it's also about understanding the depth and nuance required to achieve the desired output. Recognizing the character limit and the depth of prompts can make a substantial difference in the results obtained. Here are some characteristics of prompts that should be considered.

Some Platforms May Have Prompts Character Limitation

Many AI models, especially language models, have a maximum token or character limit for their inputs. For instance, GPT-3 has a token limit that includes both the prompt and the response. If a prompt is too lengthy, it might truncate or limit the response. Therefore, it's vital to be concise and precise with your prompts, ensuring they convey the intent without being overly verbose.

Prompts Need Depth and Context

A prompt should provide enough depth and context for the model to understand the desired outcome. For example, rather than asking "Translate," a more detailed prompt could be "Translate the following English text into French." This specificity ensures that the AI knows the source and target languages.

Differentiate between Implicit versus Explicit Prompts

Sometimes, a more extended, explicit prompt might get better results than a shorter, implicit one. For instance, "Write a detailed summary about World War II focusing on its causes" might yield more focused results than simply "Tell me about World War II."

Testing Often and Iterate

It's often beneficial to test various prompt depths—starting from concise prompts to more detailed ones—to observe which gives the best result. This iterative approach helps in refining the prompt's effectiveness.

In summary, recognizing the character and depth of prompts is crucial in leveraging AI's capabilities effectively. While being concise is necessary due to token limits, ensuring the prompt has enough depth and context is equally important to guide the AI toward the desired outcome.

Understanding a Prompt Dictionary

A prompt dictionary serves as a compendium of carefully crafted prompts tailored for specific tasks, ensuring efficient communication with AI models. This guide aims to break down the concept and benefits of a prompt dictionary.

So what is a prompt dictionary? A prompt dictionary is a curated list or database of standardized prompts, designed to achieve specific responses or actions from an AI model. Think of it as a "phrasebook" for interacting with AI.

Some advantages of using a prompt dictionary are as follows.

Consistency

One of the primary benefits of a prompt dictionary is ensuring consistency. Regardless of who is using the AI, utilizing standardized prompts from the dictionary guarantees uniformity in responses, making the AI's outputs predictable and reliable.

Optimization

Over time, as users interact with AI models, they identify which prompts yield the best results. A prompt dictionary can be continuously updated to include these optimized prompts, thereby enhancing the model's efficiency and accuracy.

Training and Onboarding

For new users unfamiliar with interacting with AI models, a prompt dictionary serves as a valuable resource. It provides them with a ready list of effective prompts, ensuring they can achieve desired results without an extended learning curve.

Flexibility

While a prompt dictionary provides standardized prompts, it can also include variations or alternative phrasing to cater to different contexts or nuances.

In essence, a prompt dictionary streamlines the process of interacting with AI models. It not only ensures consistency but also aids in achieving more accurate and meaningful responses. For anyone regularly working with AI, maintaining and regularly updating a prompt dictionary can be a game changer.

Key Factors to Consider: Context, Clarity, and Conciseness

Providing Context

Context, in the realm of AI prompt creation, is the backbone that provides a comprehensive background to any query, ensuring that the resultant answer is not just accurate but also relevant to the user's intent. The significance of context stems from several factors:

By incorporating context, the AI system is better equipped to understand the nuances of a query. For instance, the prompt "Translate this English text about medical procedures to French" offers more clarity than just "Translate this to French." The contextual information about medical procedures ensures a translation that considers medical terminologies.

A well-defined context can dramatically reduce the AI's chances of misinterpreting a prompt. If a user asks about "Java," the response could vary from programming to geography. But if the context specifies "Java programming," the AI zeroes in on the relevant topic.

Users often seek detailed, tailored responses. By understanding the context, AI can avoid generic answers. For example, "What's the weather like?" could yield a current forecast, but "What's the weather like in Seattle in December?" offers a context-specific, actionable answer.

Instead of engaging in a prolonged back-and-forth with the AI for clarifications, a contextual prompt can often lead to the desired answer in a single interaction.

In essence, context is akin to a compass for AI systems, providing the direction they need to generate insightful, accurate, and user-relevant responses. Properly harnessing context can elevate the quality of interactions and provide a smoother user experience.

Clarity

Clarity is a crucial factor in the realm of AI prompt creation. It ensures that the AI model grasps the user's intention without ambiguity, leading to more accurate and useful outputs. Here's why clarity is paramount:

Clear prompts remove any vagueness, guiding the AI toward a precise understanding. For instance, asking for "information on apple" could lead to data about the fruit or the tech company.

However, specifying "information on Apple Inc." provides clarity. An explicit prompt often means the AI can generate a response more swiftly, without the need to consider numerous possible interpretations or ask follow-up questions.

Users typically prefer a straightforward interaction where they ask a question and receive a direct answer. Clarity in the initial prompt minimizes the chances of miscommunication, making the experience seamless. Ambiguous prompts might lead AI to process vast amounts of unnecessary information before generating a response. A clear prompt ensures optimal utilization of computational resources.

When users consistently receive accurate and relevant responses due to their clear prompts, they tend to trust the AI system more. This trust is vital for widespread adoption and user satisfaction.

Clarity is foundational for effective communication, not just between humans but also between humans and AI. By ensuring that prompts are clear, users can extract the maximum utility from AI systems, ensuring productive and frustration-free interactions.

Conciseness

Conciseness in prompt creation is an art of conveying a message in the fewest words without sacrificing clarity. When interacting with AI models, brevity can be a vital tool for several reasons:

AI models, although complex, thrive on clear and succinct instructions. A concise prompt can often lead to a quicker and more accurate response because the model doesn't have to sift through superfluous information. For users, especially in business or professional settings, time is of the essence. Being able to craft a brief yet clear prompt can save valuable seconds and enhance productivity.

Shorter prompts generally require less computational power to process, ensuring that the system operates efficiently and minimizing potential costs. Every additional word in a prompt is a potential avenue for ambiguity. By keeping prompts concise, users can reduce the chances of the AI misinterpreting their request. A concise interaction reduces cognitive load for users. They don't need to craft long, complex sentences but can instead rely on short, clear prompts to get the information or action they desire.

In essence, conciseness is not just about brevity; it's about optimizing communication. It ensures that AI interactions are smooth, efficient, and free from unnecessary complexities. As AI systems continue to permeate various sectors, understanding the power of concise prompts becomes increasingly crucial for users aiming to harness the full potential of these technologies.

Some Everyday Usage Examples

The marriage of context, clarity, and conciseness can be observed in several practical scenarios when interacting with AI models. Here are some illustrative examples:

Chatbot Customer Service Poor: "Help me!"
Improved: "How do I reset my password?"
Best: "Instructions for password reset?"
Here, the last option is both concise and clear, eliminating any ambiguity about the user's intent and thereby expediting the resolution process.

Medical Diagnosis AI Poor: "I don't feel well."
Improved: "I have a fever and sore throat."
Best: "Symptoms: fever, sore throat."
The final prompt provides all necessary context and is free from extra words, which can be critical in a medical emergency.

Content Creation Poor: "Write something about climate change."
Improved: "Write a 500-word essay on climate change impacts."
Best: "500-word essay, climate change impacts."
The latter prompt is succinct but still contains all the critical information for generating the required content.

Finance Analysis Tools Poor: "Tell me about stocks."

Improved: "What's the current value of Apple stocks?"

Best: "Current value, Apple stocks?"

In finance, where real-time information is crucial, the last prompt could yield faster and more precise results.

Smart Home Devices Poor: "Can you make it cooler here?"

Improved: "Set the thermostat to 22 degrees."

Best: "Thermostat, 22 degrees."

For smart home devices that rely on voice commands, brevity and clarity can make interactions more natural and efficient.

By optimizing for context, clarity, and conciseness, users can interact more effectively with AI models across a variety of applications.

Common Mistakes to Avoid When Crafting Prompts

Crafting prompts for AI models is as much an art as it is a science. AI models continue to become more complex and versatile, and the nuances of how they interpret prompts is rapidly changing. As a result, it's increasingly essential to understand not just what works but also what doesn't. Some common pitfalls to avoid are highlighted in the following section.

Being Overly Vague Mistake: Using general or ambiguous language.

Example: "Tell me something interesting."

Consequences: The AI can return a wide variety of results, many of which might not be relevant to the user's actual intent.

Solution: Specify the domain or context, e.g. "Tell me an interesting fact about space."

Overcomplicating the Prompt Mistake: Using lengthy and complex sentences when a simpler one would do.

Example: "Can you provide me with a list of all the prime numbers that are below the number 100?"

Consequences: This can confuse the model or result in unnecessary computational processing.

Solution: "List prime numbers below 100."

Not Providing Enough Context Mistake: Leaving out key details that would guide the model's response.

Example: "Translate the following." (without mentioning the source and target language)

Consequences: The AI might make assumptions, possibly choosing a default language, or ask for further clarification, slowing the interaction.

Solution: "Translate the following from English to French."

Assuming the Model Knows the Latest Context Mistake: Assuming the AI remembers past interactions or has knowledge of recent, post-training events.

Example: "What did I ask earlier?" or "Who won the latest Oscars?"

Consequences: Models such as GPT-3 don't have memory of past interactions, and their training data might not cover the most recent events.

Solution: Always provide necessary context within the prompt or inquire about events within the model's last training data cutoff.

Using Jargon or Overly Technical Language Mistake: Assuming the AI will understand highly technical terms without context.

Example: "Explain eigenvalues." (in a non-mathematical conversation)

Consequences: The response might not be tailored to the assumed expertise level of the user.

Solution: "Explain eigenvalues in simple terms."

Ambiguous Phrasing Mistake: Using words or phrases that can be interpreted in multiple ways.

Example: "How heavy is a cricket?"

Consequences: The AI could interpret "cricket" as the sport or the insect, leading to confusing answers.

Solution: "What's the weight of an average cricket insect?"

Not Specifying the Desired Format Mistake: Not guiding the AI on how you want the answer presented.

Example: "Tell me about World War II."

Consequences: The model might provide a broad overview when you wanted a timeline or specific details about a battle.

Solution: "Provide a timeline of major events in World War II."

Neglecting to Set Boundaries Mistake: Not setting explicit guidelines, which can lead to overly verbose or out-of-scope answers.

Example: "Write about the ocean."

Consequences: The model might generate a lengthy general essay rather than focusing on a specific aspect.

Solution: "Write a short paragraph about oceanic zones."

Relying Solely on Implicit Bias Mistake: Not recognizing that AI models can have biases based on their training data.

Example: "Who is the best artist?"

Consequences: The answer can reflect cultural biases or popular opinions from the training data.

Solution: Frame questions objectively or seek data-backed answers.

Expecting Humanlike Intuition Mistake: Assuming the AI will understand human nuances, humor, or cultural references.

Example: "Explain the joke behind 'why did the chicken cross the road?'"

Consequences: While the AI can provide an explanation, it doesn't "understand" humor in the way humans do.

Solution: Understand the AI's strengths and limitations. Use it for data-driven insights rather than humanlike intuition.

Over-Relying on AI's Accuracy Mistake: Assuming every answer the AI provides is 100% correct without verifying.

Example: "Give me the complete list of symptoms for disease X."

Consequences: While AI strives for accuracy, it's not infallible and can sometimes miss nuances or recent updates in information.

Solution: Always cross-check critical information using trusted sources and don't rely solely on AI for medical or legal advice.

Not Iterating or Refining the Prompt Mistake: Accepting the first answer without trying different prompt structures.

Example: If a vague prompt doesn't provide the desired answer, some users may not refine their question.

Consequences: Settling for incomplete or not fully relevant information.

Solution: If the first response isn't satisfactory, rephrase or specify the query further.

Misinterpreting Open-Ended Responses Mistake: Asking open-ended questions and expecting a definitive answer.

Example: "What's the meaning of life?"

Consequences: Getting philosophical or generalized answers that might not meet user expectations.

Solution: Ask more focused and concrete questions to get specific answers.

Ignoring the Temperature Setting Mistake: Not adjusting the "temperature" setting (in models where this is available), which influences the randomness of the output.

Example: Keeping a high temperature for a prompt that requires a precise answer.

Consequences: Getting varied and potentially off-topic answers.

Solution: For specific, clear answers, use a lower temperature; for more creative prompts, a higher setting might be appropriate.

Assuming AI Understands Emotions Mistake: Believing that AI can empathize or understand human emotions deeply.

Example: "How do I cope with a breakup?"

Consequences: While AI can offer general advice or steps based on data, it lacks genuine human empathy.

Solution: Understand that AI responses are based on patterns and data, not genuine emotional understanding. For emotional or psychological issues, seek human support or professional counseling.

In conclusion, while AI models, especially language models, have come a long way in understanding and generating humanlike text, they're not infallible. Crafting effective prompts is a skill that requires understanding both the potential and the limitations of the AI system. By avoiding these common mistakes, users can have more productive and accurate interactions with AI models.

Diving Deeper: Structure and Nuances of Prompts

Table of Contents

Understanding Different Components of a Prompt

Introduction to the Anatomy of a Prompt

In the realm of artificial intelligence, prompts serve as catalysts, guiding AI models to produce desired outputs. However, behind this apparent simplicity of a question or statement lies an intricate anatomy that makes each prompt effective. A well-crafted prompt is more than just a question; it's a careful amalgamation of trigger words, subjects, and context, all working in concert. Trigger words set the action into motion, acting as initializers, be it to "describe," "list," or "elaborate." Following this is the subject, the essence of the inquiry, informing the AI about the central theme under discussion. But a prompt doesn't stop at mere subjects.

Modifiers and contextual embeddings refine this subject, adding layers of specificity and depth. They can narrow down broad topics, making the AI's response more relevant to a particular context or timeframe. This anatomy is further enriched by constraints that set boundaries and tone indicators that subtly guide the style or depth of the AI's response. Furthermore, as AI evolves and our interactions with it become more iterative, feedback mechanisms are integrated, allowing for a dynamic back-and-forth that refines the output.

As with any tool, understanding the anatomy of prompts is paramount. It empowers users to harness AI's potential fully, ensuring that the responses generated are not just accurate but also contextually rich and meaningful, fostering a harmonious human-AI collaboration.

The Role of Context in Shaping AI Responses

At the core of effective communication lies the essence of context, and this holds especially true for interactions with AI systems. Context in prompts acts as the guiding light, enabling AI models to navigate through vast seas of information and zero in on a specific response. Without it, AI models may deliver results that, while technically accurate, could be misplaced or lack relevance.

For instance, asking a model about the "Revolution" can lead to a myriad of responses spanning historical, technological, or even astronomical domains. But by adding context, such as "French Revolution in the 18th century," the prompt becomes anchored to a specific event, yielding a more targeted output.

Furthermore, context goes beyond just narrowing down topics. It aids in understanding nuances, sentiments, and subtexts, which are critical in areas such as sentiment analysis or cultural interpretations. By understanding the backdrop against which a query is set, AI models can adjust their tone, depth, and style of response.

Moreover, in iterative dialogues where users have a back-and-forth with the model, prior messages often provide the context for subsequent ones. Recognizing and retaining this context across interactions is crucial for the model to generate coherent and logical outputs. In essence, context is the bridge that ensures that the vast capabilities of AI are channeled appropriately, making the difference between a generic answer and one that's tailor-fitted to the user's needs.

Importance of Specificity and Clarity in Prompt Formulation

In the realm of AI, the way we communicate our inquiries is pivotal in obtaining desired outcomes. Specificity and clarity in prompt formulation are crucial determinants of this success. When prompts are vague or overly broad, the AI system faces the challenge of deciphering the user's true intent, leading to answers that might range from being mildly off-target to completely unrelated.

Specificity in a prompt acts like a compass, pointing the AI in a clear direction. For example, asking an AI, "Tell me about bears" could yield a general overview of bears. In contrast, a more specific prompt such as "Discuss the dietary habits of polar bears" directs the AI to provide focused information on that particular aspect. This nuance ensures that the user receives information that is both relevant and detailed.

Clarity, on the other hand, eliminates ambiguity. An unclear prompt can be misinterpreted, resulting in outputs that might be factually correct but contextually mismatched. For instance, a prompt such as "How high can it go?" is ambiguous. Are we discussing mountains, airplanes, stock prices, or something else? A clear prompt, such as "What's the maximum altitude commercial airplanes can reach?" leaves little room for misinterpretation.

In essence, incorporating specificity and clarity in prompt formulation is akin to fine-tuning a radio to catch a clear signal. It ensures that the vast computational prowess of AI is harnessed accurately, optimizing the quality and relevance of the generated response.

Using Temperature and Max Tokens to Guide AI Responses

Defining Temperature: How It Influences Randomness Temperature, in the realm of generative AI models, serves as a pivotal control knob for the randomness and creativity of the model's outputs. Think of it as a thermostat for the AI's "imagination"—adjusting how "hot" or "cold" its responses should be.

At its core, temperature affects the probability distribution of the model's next word choices. A higher temperature, such as values close to 1.0 or above, smoothens this distribution. This means the AI is more likely to pick words or phrases that are less common, thus resulting in more diverse and unexpected responses. It's akin to encouraging the AI to think outside the box, often leading to more creative outputs.

Conversely, a lower temperature, with values nearing 0.2 or below, narrows down the distribution, making the AI more deterministic. The model tends to produce outputs that align closely with the most probable or frequently seen sequences in its training data. Such settings are valuable when a more predictable and consistent answer is the goal.

However, it's worth noting that there are trade-offs. While higher temperatures can breed creativity, they can also lead to outputs that are less coherent or more tangential. Lower temperatures, though offering precision, might sometimes sound repetitive or overly conventional.

In essence, temperature offers a way to dial in the right balance between creativity and predictability, letting users tailor the AI's behavior to suit specific tasks and preferences.

The Significance of Max Tokens in Controlling Response Length In the domain of generative AI, the concept of max tokens is crucial for controlling the length and precision of the model's outputs. Tokens can be understood as chunks of information—often words or parts of words in text-based models. By setting a maximum limit on these tokens, users can directly influence how lengthy or concise they want the AI's response to be.

Max tokens serve multiple purposes. Firstly, they ensure that the output remains manageable and readable. In scenarios where concise answers are preferred—such as quick fact-checking or brief explanations—limiting tokens can help in getting straight to the point. Conversely, for more in-depth discussions or explorations of a topic, a higher token limit can be set.

Moreover, setting max tokens can be particularly beneficial for applications that have strict space constraints, such as generating tweets, SMS, or producing content for specific platforms with character limits. It ensures the content fits the desired space without manual truncation or editing.

However, it's essential to strike a balance. Setting the token limit too low might result in outputs that are overly terse or might not fully address the query. On the other hand, too high a limit, especially without a corresponding context, might lead to verbose answers that drift off-topic.

In conclusion, max tokens play a pivotal role in shaping the AI's output to match user expectations, ensuring the generated content is both relevant in content and appropriate in length for its intended purpose.

Practical Examples: Adjusting Temperature and Max Tokens for Desired Outcomes In the realm of AI-powered content generation, both temperature and max tokens serve as influential dials, significantly influencing the nature and size of the output. Here's a glimpse into how tweaking these parameters can lead to varied results in practical scenarios:

Story generation: Imagine an AI is tasked to generate a sci-fi plot. A higher temperature setting, say 0.8, could yield more unpredictable and creative twists, while a lower setting, around 0.2, would provide a more consistent and linear narration. If we wish for a short plot summary, setting max tokens to 50 might suffice, but for a detailed storyline, we might raise it to 500 or more.

Customer support: When using AI for instant chat support, clarity and brevity are key. A temperature close to 0.2 ensures accurate and consistent responses. Max tokens might be limited to 100 to ensure answers are concise.

Content ideation: For brainstorming blog topics, a higher temperature might encourage diverse and out-of-the-box suggestions. However, using a max token limit of 10–15 ensures that suggestions remain short and topic-focused.

Research summaries: For summarizing research papers, a balance is needed. A mid-range temperature such as 0.5 ensures a mix of creativity and adherence to the source. Setting max tokens to 300 might offer comprehensive abstracts.

In essence, the interplay of temperature and max tokens allows users to mold AI outputs, tailoring them for specific applications and desired characteristics. Mastering these parameters unlocks the potential to harness AI's capabilities effectively across diverse tasks.

Balancing Creativity and Control in Your Prompts

The Trade-Off between Open-ended and Specific Prompts Prompts serve as the bridge between human intent and AI-generated content. Their structure can significantly influence the nature of AI responses.

At the core of crafting prompts lies a fundamental trade-off: the continuum between open-endedness and specificity.

Open-ended prompts: These prompts are broad, allowing the AI to tap into its extensive knowledge and creative capabilities. An open-ended prompt, such as "Write about the universe," can yield a plethora of outcomes, from a poetic contemplation of the stars to an in-depth scientific exposition. The advantage is the potential for diverse, unexpected insights. However, the downside is unpredictability; the response might not always align with the user's intent.

Specific prompts: On the other side are precise, well-defined prompts that channel the AI's response in a particular direction. For instance, "List the main types of galaxies in the universe" narrows down the expected answer. The benefit here is control; the user is more likely to receive a focused and relevant response. The trade-off, however, is potentially limiting the AI's creative range.

Finding the middle ground: The key is to determine the desired outcome. If exploration and creativity are the goals, leaning toward open-endedness makes sense. Conversely, for tasks demanding accuracy and precision, specificity is paramount.

In essence, understanding this trade-off is crucial for harnessing the true potential of AI. It's about guiding the AI, not just instructing it, ensuring a harmonious blend of creativity and control.

Techniques to Harness AI's Creativity without Losing Focus AI's power lies in its vast knowledge and ability to generate creative content. However, unleashing its full creative potential can sometimes lead to results that meander or miss the mark. Striking a balance requires techniques that guide AI's creativity toward a focused outcome.

Iterative refinement: Start with an open-ended prompt to explore AI's creative spectrum. Based on its response, iteratively refine the prompt to channel creativity in the desired direction. This approach leverages the best of both worlds, combining exploratory brainstorming with guided specifics.

Prompt layering: Compound prompts can be highly effective. Begin with a broad theme, followed by specific questions or directions. For instance, "Write a story about space. Ensure it involves a human astronaut and an alien friendship."

Use analogies: Analogies can guide AI's thinking by setting a familiar context. For instance, "Describe the concept of quantum physics as if explaining it to a five-year-old" harnesses AI's knowledge but within a controlled, simple framework.

Set boundaries with keywords: While you might want creative input, certain boundaries can keep the content on track. "Write a poem about nature, avoiding mention of the sea or mountains" sets clear exclusions, ensuring a more novel response.

Feedback loop: Use AI's own output as feedback. If a generated piece is too broad, use it as a base, and prompt the AI to focus or expand on a specific section.

In essence, while AI's creativity is a formidable asset, steering it with purposeful techniques ensures that its generative prowess is both inventive and on point.

Case Studies: Successful Balance in Real-World Applications

Content Marketing

Scenario: A tech company wanted to generate articles about AI trends.

Approach: They prompted the AI with "Describe the top five AI trends in 2022, ensuring each trend is explained in layman's terms."

Result: The AI produced a listicle that was both insightful, catering to industry professionals, and accessible to general readers. The specificity of the prompt ensured relevance, while the directive for simplicity ensured wide appeal.

Scriptwriting

Scenario: A filmmaker sought AI's help to brainstorm a sci-fi movie plot.

Approach: Using the prompt "Create a sci-fi plot set in a future where humans can transfer memories. Focus on a love story."

Result: The AI spun a unique tale of two lovers reliving their past, blending futuristic elements with emotional depth. The love story focus kept the narrative tight and relatable.

Fashion Design

Scenario: A fashion start-up aimed to design a new line of sustainable summer wear.

Approach: They input, "Design outfits for summer using sustainable materials. Think beach meets city."

Result: The AI sketched concepts that beautifully merged casual beach vibes with urban sophistication, all rooted in sustainability. The dual-theme prompt led to hybrid designs that were innovative yet marketable.

Music Composition

Scenario: A game developer needed background scores for a forest level.

Approach: "Compose a melody that captures a dense, magical forest at dawn."

Result: The AI generated a piece evoking early morning chirps and mystical undertones, enhancing the gaming experience.

These cases underline the power of carefully crafted prompts. When clarity meets creativity, AI's output can be both groundbreaking and precisely targeted.

The Art of Iterative Prompting

Importance of Refining Prompts Based on AI Feedback In the realm of generative AI, the relationship between the user and the model is one of continuous dialogue. It's a dynamic dance where input and output shape one another. Central to this interaction is the importance of refining prompts based on the feedback received from the AI, a process akin to tuning an instrument to produce the desired sound.

Firstly, AI models, even the most sophisticated ones, are not omniscient entities. They generate responses based on patterns they've learned from vast amounts of data. When a user poses a question or gives a command, the AI offers what it deems to be the most appropriate response. However, this initial answer might not always align with the user's intent or expectations.

This is where refining prompts becomes crucial. By adjusting the language, specifying details, or rephrasing the request, users can glean more targeted responses from the AI. For instance, if an AI-generated story lacks a dramatic climax, the user can prompt it to enhance tension or introduce a twist.

Moreover, iterative prompting serves as an essential feedback mechanism for the model itself. The more specific and refined the prompts, the more the AI learns and adapts to user preferences, creating a tailored user experience.

In essence, prompt refinement, based on AI feedback, transforms the AI-user interaction from a static Q&A session to a dynamic, co-creative process. It underscores the importance of treating AI not just as a tool but as a collaborator, where feedback and adaptation lead to richer, more accurate, and more valuable outcomes.

Strategies for Iterative Improvement in Prompting

The art of iterative prompting is akin to fine-tuning a musical instrument. Each adjustment brings you closer to the desired harmony. As AI models continue to evolve, understanding how to iteratively improve your prompts can dramatically enhance the quality of responses. Here are some strategies to ensure optimal results:

Start broad and then narrow down: Begin with a generic prompt to gauge the AI's initial response. From there, adjust your prompts to be more specific based on the nuances of the output you're seeking.

Feedback loops: After each response, identify areas of divergence from your desired outcome. Use these insights to rephrase or add specificity to your subsequent prompts.

Varying language: If you're not getting the desired results, rephrase your prompt using different vocabulary or sentence structures. The way you phrase a question can drastically alter the AI's response.

Contextual embedding: Supply the AI with context. If you're looking for a continuation of a story or idea, provide a brief summary or the last known data point to guide the AI's generation process.

Prompt splitting: If a prompt contains multiple components, try breaking it down into individual parts. Seek answers to each part separately, and then piece them together.

Experiment with parameters: If the platform allows, play with settings such as temperature for randomness and max tokens for length. Sometimes, the key lies in these subtle adjustments.

Learn from failures: Not every response will be perfect. Treat failed attempts as learning opportunities, analyzing why the AI might have responded in a certain way.

Iterative prompting is a dynamic process. It requires patience, keen observation, and a willingness to engage in a feedback-driven dialogue with the AI. With these strategies, users can harness the full potential of generative models to meet their specific needs.

Common Pitfalls and How to Avoid Them

When fine-tuning prompts for optimal AI outputs, users often navigate a trial-and-error process. Recognizing common pitfalls can pave the way for a more effective prompting approach. Here are some of the frequent challenges encountered and strategies to mitigate them:

Vagueness: Broad or unspecific prompts can lead the AI to produce generalized responses. To counteract this, always be clear and concise about your requirements. For instance, instead of "Tell me about climate," specify "Explain the causes of climate change."

Overloading information: While details can be helpful, overwhelming a prompt with unnecessary information might confuse the AI or deviate its focus. Keep prompts relevant and to the point.

Being rigid: Sticking to one phrasing or structure can limit the AI's potential. If you're not getting the desired output, rephrase the question or approach the topic from a different angle.

Over-reliance on parameters: While adjusting parameters such as temperature can be beneficial, they're not a panacea. Rely on them in tandem with well-crafted prompts, not as a primary solution.

Ignoring context: Especially in a chain of queries, maintaining context can be pivotal. Always ensure the AI has the necessary background, especially if building on previous responses.

Assuming AI knows best: Remember that AI models, regardless of their complexity, don't possess human intuition. If a response seems off, don't hesitate to re-prompt or seek further clarification.

Skipping review: Always review and reflect on AI-generated content. Continuous assessment helps identify areas for prompt refinement.

In summary, effective iterative prompting hinges on clarity, flexibility, and continuous learning. Being aware of these pitfalls and proactively addressing them can greatly enhance the user's experience with generative AI models.

Advanced Prompting Techniques

Using External Knowledge and Context to Enhance Prompts In the realm of advanced prompting techniques, the integration of external knowledge and contextual understanding plays a pivotal role in eliciting richer and more relevant responses from AI models.

Incorporating domain-specific lexicons: By introducing specific jargon or terminology pertinent to a particular field, prompts can guide AI to produce answers that resonate with expert audiences. For instance, in a medical context, using terms such as angioplasty or hemoglobin levels can steer the response toward a more clinical tone.

Historical and temporal context: Embedding a time-based context, such as referencing an event or era, can help in obtaining time-specific information. A prompt such as "Discuss nuclear energy's popularity post-Chernobyl disaster" brings forth responses grounded in the aftermath of that event.

Cultural and geographical nuances: Introducing cultural or regional elements can shape the AI's output to be more aligned with local customs, values, or events. A prompt about "Diwali celebrations in North India" would generate a culturally nuanced response about this specific region.

Reference to authoritative sources: By suggesting that the AI should base its response on principles from a specific book, research paper, or expert, the user can achieve answers that mirror the tone or viewpoint of these references.

Scenario-based contextualization: Providing a hypothetical situation can help in drawing out more imaginative or solution-oriented responses from the AI. For instance, "Imagine a world without fossil fuels; how would transportation evolve?"

Incorporating external knowledge and diverse contexts into prompts not only refines the accuracy of AI-generated content but also tailors responses to resonate more profoundly with the intended audience or objective.

Incorporating User Feedback for Dynamic Prompting Harnessing user feedback is an advanced and vital technique to refine and enhance the effectiveness of prompts in AI-driven systems. Dynamic prompting, built on the feedback loop principle, allows the AI system to adapt and evolve in real time based on user interactions and reactions.

Immediate refinement: By allowing users to flag inaccurate or unsatisfactory outputs, systems can instantly reprocess the information and provide a better response. This immediate feedback mechanism ensures a more user-centric model of content generation.

Learning from mistakes: Over time, consistent feedback allows AI models to identify recurring errors or biases in their responses, leading to reduced errors in future interactions.

Customization: In applications such as chatbots or customer support AI, dynamic prompting helps in tailoring responses based on the user's feedback. For instance, if a user prefers detailed explanations over succinct ones, future interactions could adapt to this preference.

Quality control: Platforms that use AI for content generation or decision-making can have user rating systems. Responses that receive higher ratings can influence the model to produce similar outputs in the future.

Iterative training: Accumulated feedback can be looped back into the training data. This continuous model refinement ensures that the AI system remains updated with the latest user preferences and external data changes.

Incorporating user feedback for dynamic prompting fosters a symbiotic relationship between the user and the AI system. As users strive for more accurate and relevant information, AI systems, through dynamic prompting, become more attuned to user needs, ensuring more meaningful and precise interactions.

Exploring the Future: Evolving Trends in AI Prompting The landscape of AI prompting is rapidly evolving, with emerging trends promising to redefine the interactivity and intelligence of AI systems. Here's a glimpse into the future of AI prompting:

Personalized prompting: With the increasing integration of AI in daily tech, systems will generate prompts tailored to individual users, understanding their preferences, past interactions, and contextual needs. This personal touch will revolutionize user experience.

Multimodal prompts: Future AI models will not just rely on text. They will incorporate multimodal inputs such as images, voice, or even gestures to generate a response, enhancing the richness of human-AI interaction.

Adaptive learning: Advanced AI models will self-adjust their prompting strategies based on real-time feedback. Rather than waiting for large datasets for training, these models will make incremental adjustments continuously.

Proactive prompting: AI systems will anticipate user needs and provide prompts even before a direct query, thanks to predictive analytics and deep learning insights.

Ethical and bias checks: With the increasing awareness of AI biases, there will be a robust mechanism to ensure that prompts don't lead to biased or unethical outputs. Systems will be trained to recognize and avoid potentially harmful content.

Integration with augmented reality (AR) and virtual reality (VR): As AR and VR technologies mature, AI prompts will play a pivotal role in shaping immersive experiences, guiding users contextually within these virtual environments.

As we venture deeper into the AI-driven future, the art and science of prompting will become even more crucial. It holds the key to making AI tools not just smart but also intuitive, empathetic, and ethically sound.

Prompt Engineering across Industry

Table of Contents

Mastering Creative Writing Use Cases for Content Creators

Prompt engineering, an integral component of the burgeoning AI domain, offers novel possibilities to the world of creative writing. By effectively utilizing and guiding AI models, content creators are exploring fresh avenues, enhancing their productivity, and presenting diversified content. Here's a succinct exploration of this phenomenon.

Brainstorming and Ideation The starting point of any writing venture is often the most challenging. With AI, writers can input generic themes or topics, receiving a myriad of suggestions ranging from intricate plot twists to expansive thematic explorations. For instance, inputting "a mystery set in Victorian London" could spawn ideas about characters, potential crimes, or even intertwining subplots.

Combating Writer's Block One of the most dreaded challenges for writers is the notorious *writer's block*. Through prompt engineering, when faced with a stalemate, writers can gain a nudge in the right direction. By inputting the last coherent thought or plot point, AI can suggest potential continuations or shifts, restoring the creative flow.

Adaptive Storytelling Modern readers often seek dynamic content that evolves with real-time events or trends. Through AI prompts, writers can mold narratives that integrate current happenings, creating stories that resonate deeply with the present zeitgeist.

Diverse Stylistic Outputs Every writer aspires to cultivate a unique voice, but there's also a charm in emulating classic styles. By inputting prompts related to specific authors or literary epochs, AI can produce content reminiscent of those iconic voices, be it the brevity of Hemingway or the descriptive allure of Tolkien.

Feedback Loop In today's digital age, reader feedback is instantaneous. Writers can prompt AI systems to analyze feedback, discerning prevalent sentiments or critiques. This real-time analysis allows for immediate content refinement, fostering a more engaged readership.

Case Study: The Digital Novelist Consider The Digital Novelist, a platform that revolutionized serialized fiction. Writers input their story's primary plot, characters, and setting. The AI, based on daily global news and real-time reader feedback, suggested twists and turns, keeping the narrative perpetually fresh and relevant. The platform saw a 200% increase in daily readership, with writers praising the tool for its invaluable assistance in maintaining consistency while navigating the fluidity of real-time events.

Prompt engineering, far from undermining the writer's role, acts as an amplifier of their creativity. It's a harmonious blend of human imagination and machine precision, ushering in a new era for content creators where boundaries are constantly redefined and possibilities are endless.

Transforming Operations: Business Applications of Prompt Engineering

In the ever-evolving landscape of business, adaptability and innovation are paramount. With the rise of AI, prompt engineering emerges as a game changer, offering businesses a competitive edge through nuanced, AI-guided solutions. Let's delve into how prompt engineering is reshaping various business operations:

Supply chain optimization: Efficient supply chain management hinges on predicting and navigating countless variables, from supplier reliability to transit delays. By crafting precise prompts, businesses can gain AI-driven insights into optimal inventory levels, potential disruptions, or best-suited suppliers, ensuring smoother operations.

Human resources management: The recruitment process can be arduous, sifting through countless applications to identify potential fits. HR professionals can use prompts to guide AI in screening applications, analyzing responses for cultural fit, or even automating preliminary interview questions, streamlining the hiring process.

Customer service enhancement: Modern businesses prioritize customer experience. AI chatbots, driven by well-engineered prompts, can handle customer queries, complaints, or feedback efficiently. They can be tailored to align with a company's ethos, ensuring consistent and timely customer interactions.

Market analysis: Understanding market trends, competitor strategies, or consumer preferences is pivotal. Through prompt engineering, businesses can instruct AI to dissect vast data troves, generating insights into emerging market trends, potential investment areas, or areas needing innovation.

Financial forecasting: For businesses, financial foresight is crucial. By feeding historical financial data and prompting AI with specific queries, businesses can gain forecasts on sales, profitability, or potential financial risks, aiding in informed decision-making.

Case Study: NexaRetail's Revolution Consider the meteoric rise of NexaRetail, a midsize retailer. Recognizing the potential of prompt engineering, they integrated AI into their operations. Inventory management was optimized through prompts that analyzed sales data, predicting stock requirements. Their customer service chatbots, guided by meticulously crafted prompts, ensured 24/7 customer engagement, driving up satisfaction scores. Furthermore, financial prompts provided insights into seasonal sales trends, allowing NexaRetail to adjust marketing strategies dynamically. Within a year, they reported a 30% increase in operational efficiency and a 20% boost in sales.

Prompt engineering is not just a technological advancement; it's a business revolution. By bridging the gap between vast data reservoirs and actionable insights, it provides businesses a sharper, more efficient operational lens. As companies recognize its potential, we're witnessing a paradigm shift toward data-driven, AI-assisted decision-making, ensuring businesses remain agile, efficient, and perpetually ahead of the curve.

Revolutionizing Education Classroom Applications of Prompt Engineering

As the digital age sweeps across industries, education stands at the cusp of a monumental shift. AI, combined with the power of prompt engineering, is ushering in an era of personalized, dynamic, and interactive learning. This intersection of technology and pedagogy is transforming classrooms and redefining the educational experience.

Personalized Learning Paths Every student is unique, with varied learning speeds and preferences. By crafting precise prompts, educators can utilize AI to generate individualized lesson plans or resources tailored to each student's strengths and areas of improvement, ensuring no one is left behind.

Interactive Study Materials Traditional texts are being replaced with dynamic content. Prompts can guide AI to convert complex topics into engaging formats such as animations, simulations, or interactive quizzes, enhancing comprehension and retention.

Language Learning Assistance Learning a new language can be daunting. With the aid of prompt engineering, AI-driven platforms can generate practice exercises, correct pronunciations, or even simulate conversational scenarios, making language acquisition more natural and effective.

Real-Time Feedback Immediate feedback is crucial for learning. Educators can craft prompts for AI systems to assess assignments or tests, providing students with instant, detailed feedback on their performance, highlighting areas for further focus.

Classroom Management From tracking attendance to assessing participation, AI-driven systems, guided by specific prompts, can assist educators in efficiently managing classroom tasks, allowing them to dedicate more time to instruction.

Augmented Reality (AR) Integration AR has vast educational potential. Prompts can guide AI to superimpose historical data on real-world objects or simulate scientific experiments in 3D, providing students with immersive learning experiences.

Case Study: EduFutura's Transformation EduFutura, an ed-tech start-up, harnessed the power of prompt engineering to reshape its digital learning platform. They integrated AI-driven chatbots, which, when prompted, offered students study resources based on their past performance and current queries. Additionally, their virtual science lab, guided by AI prompts, simulated experiments, allowing students to explore and learn in a risk-free environment. Within months, Edu-Futura witnessed a 50% increase in active users, with students reporting enhanced understanding and engagement.

Education, at its core, is about nurturing minds and preparing them for the future. With the integration of prompt engineering, the boundaries of traditional pedagogy are expanding, providing students with a more holistic, engaging, and tailored learning experience. As educators and institutions recognize and harness its potential, we stand on the brink of an educational renaissance, where learning is not just about rote but about exploration, interaction, and personal growth.

Power to the Press Journalism and Media Use Cases

In an age of information deluge, journalism and media organizations face challenges in producing accurate, timely, and compelling content. With the advent of AI, and especially prompt engineering, the media landscape is undergoing a transformative shift, leveraging technology for enhanced storytelling and informed reportage.

Speedy Reporting In a world where real-time updates are critical, journalists can utilize prompts to extract quick summaries or important highlights from vast datasets or events, ensuring timely news updates without compromising accuracy.

Data Journalism Analyzing complex datasets to derive meaningful stories is pivotal in modern journalism. By crafting specific prompts, journalists can guide AI to spot trends, anomalies, or patterns in data, enabling deep, investigative pieces.

Content Personalization To cater to diverse audiences, media houses can employ prompt engineering to curate content based on user preferences, reading history, or even regional events, ensuring readers always find relevance in what they consume.

Automated Transcription and Translation Covering global events requires rapid transcription and translation. AI systems, with the right prompts, can swiftly transcribe interviews or translate foreign content, allowing journalists to focus on analysis and content creation.

Fact-Checking In an era where misinformation is rampant, ensuring content credibility is paramount. Through prompt engineering, AI can be directed to cross-reference statements, claims, or data against trusted sources, bolstering the integrity of the report.

Visual Storytelling The adage "a picture is worth a thousand words" holds. Media organizations can use prompts to guide AI in suggesting or creating optimal graphics, infographics, or even interactive visuals that complement and elevate the narrative.

Case Study: NewsNet's Evolution NewsNet, a global news agency, integrated prompt engineering to optimize its content delivery. Utilizing AI-driven prompts, their platform curated news based on readers' interests and past consumption patterns, increasing user engagement by 40%. Additionally, their investigative team, leveraging data-journalism prompts, unveiled significant socioeconomic trends, winning accolades for in-depth reporting. Their AI-driven fact-checker, guided by meticulously structured prompts, ensured that real-time news remained free from misinformation, establishing NewsNet as a beacon of trustworthy journalism in turbulent times.

Prompt engineering is not just a technological tool; for journalism and media, it's a beacon guiding them through the complexities of modern reportage. By ensuring speed, accuracy, personalization, and depth, this fusion of technology and storytelling is setting new benchmarks in journalism. As media houses globally integrate prompt engineering, readers are assured content that isn't just timely and relevant but also rich, insightful, and credible.

Streamlining Research Applications for Academics and Market Analysts

Research, whether academic or market-driven, stands as the backbone of informed decision-making and progress. In today's data-saturated environment, extracting meaningful insights demands more than just analytical skills. Here's where prompt engineering plays a pivotal role, offering a confluence of human expertise and machine efficiency.

Literature Reviews Academics spend countless hours perusing previous works for their studies. By crafting tailored prompts, researchers can guide AI to highlight relevant studies, seminal works, or pivotal data points, significantly reducing the time taken for literature reviews.

Data Analysis Market analysts, inundated with vast datasets, can use prompt engineering to extract specific trends, consumer behaviors, or anomalies. Whether it's identifying sales patterns or tracking market shifts, a well-engineered prompt can provide insights in seconds.

Predictive Modeling For academics and analysts predicting future trends or outcomes based on historical data, AI, when given the right prompt, can create predictive models, offering glimpses into potential future scenarios or market movements.

Survey Analysis Gleaning insights from extensive surveys can be tedious. With appropriate prompts, AI can swiftly categorize open-ended responses, identify sentiment trends, or even spotlight outliers, streamlining the analysis process.

Content Creation Drafting reports, papers, or market summaries demands precision. Researchers can guide AI through prompts to structure their findings, ensuring content is both coherent and relevant.

Real-Time Tracking Market analysts often need real-time data tracking for volatile sectors. Prompt engineering can be utilized to direct AI systems to provide real-time alerts, summaries, or deep dives as market conditions evolve.

Case Study: ScholarSphere's Integration ScholarSphere, a research platform, harnessed prompt engineering to redefine its user experience. Academics, by inputting specific prompts, could swiftly locate relevant literature, significantly reducing their groundwork time. Market analysts, on the other hand, used the platform to extract real-time market insights, especially during critical financial events, ensuring they were always a step ahead in their analyses. The platform also facilitated researchers in creating structured drafts for their findings, blending human expertise with AI's efficiency. Within a year of these integrations, ScholarSphere saw a 70% increase in its active user base and became a staple in academic and market research circuits.

Prompt engineering, in the realm of research, is a transformative tool. It enhances efficiency, ensures depth, and provides a refined structure to the research process. As academics and market analysts embrace this synergy of human acumen and AI, research outcomes become more precise, relevant, and timely, setting new standards for excellence in the world of investigation and analysis.

Navigating the Legal Landscape Prompt Engineering in Law

The legal profession, anchored in vast archives of case laws, statutes, and intricate nuances, stands at an exciting crossroads with the advent of AI and prompt engineering. Through adept application, attorneys, paralegals, and law firms are discovering newfound efficiencies and enhanced accuracy.

Legal Research Scouring through expansive legal databases for relevant case laws, statutes, or legal commentaries is a cornerstone of legal practice. Prompt engineering enables professionals to quickly pinpoint specific cases, judgments, or references, considerably reducing research hours.

Contract Analysis Contracts are often riddled with complex legalese and clauses. With well-defined prompts, AI can assist in identifying key terms, potential pitfalls, or even suggest standardized clauses, ensuring contracts are both sound and comprehensive.

Litigation Prediction Lawyers can leverage prompt engineering to estimate the potential outcomes of a case based on historical judgments, enabling clients to make informed decisions regarding litigation.

Document Automation Routine legal documents, such as wills, lease agreements, or even simple contracts, can be swiftly generated using AI systems directed by specific prompts, ensuring consistency and saving valuable time.

Legal Discovery During litigation, discovery processes involve sifting through vast volumes of data. Tailored prompts guide AI in extracting pertinent information, documents, or communications, streamlining the often-tedious discovery phase.

Real-Time Legal Updates Laws and regulations evolve constantly. Through prompt engineering, legal professionals can receive real-time updates on amended laws, new case judgments, or evolving legal standards, keeping them perpetually abreast of the dynamic legal landscape.

Case Study: LawScape's Transformation LawScape, a cutting-edge legal tech platform, integrated prompt engineering to revolutionize its offerings. Lawyers, upon inputting specific legal queries, received curated case law references, drastically enhancing research efficiency. Their contract analysis tool, driven by prompts, flagged potential issues and suggested optimal clauses, becoming a staple for contract drafting. Additionally, the platform's litigation predictor, using historical judgments and AI-guided prompts, offered attorneys a statistical insight into potential case outcomes. This multipronged integration led to LawScape's adoption by over 50% of top-tier law firms within two years, marking a watershed moment in legal tech.

Prompt engineering, within the legal realm, isn't merely a technological addition; it's a paradigm shift. By bridging the chasm between exhaustive legal archives and immediate, precise insights, it's enabling the legal fraternity to serve justice more efficiently and accurately. As the symbiotic relationship between the law and technology deepens, prompt engineering stands as a testament to the future of legal practice—sharper, faster, and more precise.

Prompt Engineering in Health Care: Reinventing Patient Communication and Data Analysis

Health care, an industry pivotal to human well-being, thrives on accuracy, timeliness, and effective communication. The convergence of AI with prompt engineering is opening doors to a more streamlined, personalized, and data-driven approach in patient care and medical research.

Patient Interaction For health care providers, effective communication is crucial. AI-driven chatbots, equipped with tailored prompts, can answer patient queries, provide medication reminders, or even assist in appointment bookings, ensuring continuous and efficient patient engagement.

Medical Data Analysis Prompt engineering can guide AI systems to swiftly analyze intricate medical data, from patient records to lab results. Whether it's spotting anomalies in an MRI scan or predicting patient health trajectories, precise prompts ensure in-depth analysis.

Personalized Treatment Plans By processing patient histories and current health metrics, AI, through well-engineered prompts, can suggest personalized treatment paths, enhancing the effectiveness of medical interventions.

Real-Time Monitoring Wearable health tech, when integrated with AI-driven prompts, can offer real-time health updates. Be it flagging irregular heart rhythms or monitoring insulin levels, prompt engineering ensures timely alerts and interventions.

Medical Research For researchers delving into complex medical datasets, AI, guided by specific prompts, can identify patterns, correlations, or potential areas of study, accelerating the pace of medical innovations.

Telehealth Assistance In the era of remote consultations, prompt-guided AI systems can assist health care professionals by providing real-time data analysis, medication suggestions, or even gauging patient sentiment during virtual interactions.

Case Study: HealthLink's Integration HealthLink, a prominent digital health care platform, integrated prompt engineering to optimize its services. Their AI-driven chatbots, using specific prompts, addressed patient queries 24/7, reducing hospital hotline traffic by 40%. In telehealth sessions, their AI system, guided by prompts, provided doctors with real-time patient data insights, enhancing the quality of virtual consultations. Moreover, their research wing utilized prompt-engineered AI to sift through vast datasets, identifying potential correlations between lifestyle factors and chronic illnesses, leading to several groundbreaking papers.

The melding of prompt engineering with health care is not just transformational—it's lifesaving. By ensuring rapid data analysis, personalized patient care, and efficient communication, it reinforces the very ethos of health care–patient-centric, data-driven, and ever-evolving. As medical professionals globally harness this synergy, health care is set to become more responsive, precise, and patient-friendly, heralding a new era of medical excellence.

Entertainment Reimagined: Applications in Music, Film, and More

The world of entertainment, characterized by creativity and innovation, finds a dynamic ally in prompt engineering. As AI delves deeper into the arts, it's evident that the fusion of human creativity with machine precision can reinvent entertainment across its spectrum.

Scriptwriting and Storyboarding Harnessing AI can help screenwriters and directors craft compelling story arcs. With meticulously designed prompts, AI can suggest plot twists, character developments, or even thematic elements, enriching narratives and broadening creative horizons.

Music Composition Prompt engineering aids musicians in crafting melodies, harmonies, or even lyrics. By feeding AI systems musical preferences or desired moods, artists can receive unique compositional suggestions, leading to collaborative symphonies between man and machine.

Personalized Entertainment Experiences For OTT platforms or music streaming services, tailoring content recommendations for users is paramount. With apt prompts, these platforms can provide users with bespoke entertainment playlists, enhancing user engagement and satisfaction.

Film Editing Editors, inundated with hours of footage, can employ AI-driven prompts to highlight key sequences, optimal transitions, or even detect continuity errors, streamlining the post-production process.

Virtual Reality (VR) and Augmented Reality (AR) The realms of VR and AR thrive on immersive experiences. Prompt engineering can guide AI systems to customize virtual environments, be it for gaming, simulations, or experiential cinema, ensuring users always receive a unique experience.

Talent Scouting For talent agencies or production houses, sifting through a sea of portfolios can be daunting. AI, armed with specific prompts, can identify potential outstanding talents, be it for acting, singing, or other performing arts.

Case Study: EchoTunes' Evolution EchoTunes, a digital music platform, leveraged prompt engineering to redefine music creation and consumption. Their AI-driven studio suite, powered by artist-generated prompts, suggested melody lines, beat patterns, or lyrical themes, making music creation a collaborative venture. On the user end, prompts based on mood, activity, or past listening habits curated bespoke playlists. Within months, EchoTunes became the go-to platform for both music creators and aficionados, marking a paradigm shift in the music industry.

Entertainment, an industry rooted in evoking emotions, finds in prompt engineering a tool that magnifies its reach and depth. By melding creative passions with AI-driven insights, the world of music, film, and arts is witnessing a renaissance. As artists and creators globally harness this synergy, audiences are treated to content that is not just entertaining but also deeply personalized and constantly evolving.

Building Smarter Cities: Urban Planning and Public Policy Applications

Modern urban environments, brimming with complexities, necessitate solutions that are both efficient and sustainable. With the merging of prompt engineering and AI, urban planning and public policy are undergoing significant enhancements, pushing the envelope for smarter cities.

Traffic Management and Infrastructure Navigating congested city streets is a daily challenge. By utilizing prompt-engineered AI systems, cities can analyze traffic patterns, predict congestion points, and suggest optimization strategies, ensuring smoother commutes and reduced emissions.

Sustainable Urban Planning For planners envisioning green cities, AI can be prompted to design layouts optimizing sunlight, green spaces, and natural ventilation, contributing to eco-friendly urban habitats.

Public Safety and Security Prompt-driven AI can assist in real-time surveillance, spotting anomalies or potential security threats, ensuring public areas remain safe and secure.

Resource Management Whether it's water supply, waste management, or power distribution, AI systems with precise prompts can offer insights into optimal resource allocation, ensuring sustainability and reduced wastage.

Public Policy Decision-Making Lawmakers can harness AI to analyze public sentiment, economic indicators, or societal trends. With well-crafted prompts, these analyses can guide policy decisions that are data-backed and resonate with the populace.

Urban Data Visualization City administrators can employ prompt-engineered AI to create comprehensive visual representations of urban data, be it demographic distributions, infrastructural statuses, or resource consumptions, aiding in informed decision-making.

Case Study: MetroVille's Transformation MetroVille, a burgeoning urban center, tapped into the power of prompt engineering to address its growing pains. Tasked with managing rising traffic, a dedicated AI system, driven by traffic-centric prompts, suggested alternate transit modes and routes, leading to a 30% reduction in peak-time congestion. Similarly, their urban planning committee, through AI-generated visual prompts, identified areas for green space expansions, resulting in a citywide park enhancement initiative. On the policy front, sentiment analysis prompts helped lawmakers gauge public opinion on proposed regulations, ensuring policies were both effective and popular. Within a few years, MetroVille transformed from a congested urban space to a model of smart, sustainable city living.

The journey to creating smart cities, centered on the well-being and progress of their inhabitants, is bolstered by prompt engineering. This synergy of human expertise and machine intelligence ensures urban spaces are not only efficient but also inclusive, sustainable, and prepared for future challenges. As cities globally embrace this confluence, urban living is set to experience a transformation like never before, marking a new epoch of cityscapes that truly cater to their denizens.

Making Markets Prompt Engineering in Finance and Economics

In the rapidly evolving world of finance and economics, prompt engineering melds with AI to offer a paradigm shift in decision-making, risk management, and market analysis. This alliance delivers robust financial models and actionable economic insights, driving industries toward precision and predictability.

Financial Forecasting Traders and financial institutions harness AI, guided by carefully crafted prompts, to predict market movements. By analyzing historical data and current market conditions, AI provides insights into potential market trends, offering traders an edge.

Risk Management Banks and insurance companies can utilize prompt-driven AI systems to assess the creditworthiness of clients or evaluate insurance claims, ensuring decisions are both rapid and data-backed.

Algorithmic Trading Through precise prompts, AI-driven algorithms can execute high-frequency trades at speeds unimaginable to humans, capitalizing on minuscule market fluctuations and generating profits.

Economic Trend Analysis Economists employ AI models, guided by pertinent prompts, to dissect complex datasets, unveiling underlying economic patterns, growth drivers, or potential recession indicators.

Personalized Financial Products Financial institutions can tailor products, from loan offers to investment portfolios, based on individual client profiles. With the aid of prompt engineering, AI systems can suggest the most fitting financial solutions for customers.

Real-Time Financial Monitoring For regulatory bodies overseeing financial markets, prompt-guided AI can provide real-time alerts for suspicious trading activities, ensuring market integrity and preempting potential financial crises.

Case Study: FinTechPro's Mastery FinTechPro, a leading financial technology firm, employed prompt engineering to elevate its suite of services. Their AI-driven platform, when fed with specific market prompts, generated real-time financial forecasts, becoming indispensable to traders worldwide. Moreover, their personalized financial advisory, powered by AI's prompt responses, offered clients bespoke investment strategies, leading to better financial returns. On the regulatory front, their AI system, equipped with monitoring prompts, swiftly flagged irregular market activities, fortifying market robustness. Within just two years, FinTechPro's innovations led it to dominate global financial tech charts.

The integration of prompt engineering in finance and economics heralds an age of data-driven decision-making, proactive risk aversion, and bespoke financial solutions. This synergy amplifies the acumen of financial professionals, enabling markets to function with heightened efficiency and resilience. As the global financial landscape grapples with uncertainties, the partnership of human expertise and AI-driven prompts stands as a beacon, guiding the industry toward stability, growth, and unwavering trust.

Practical Guide to Prompt Engineering

Table of Contents

Step-by-Step Guide to Crafting Your First Prompt

Crafting effective prompts for a generative AI system is pivotal in guiding its responses and ensuring meaningful and accurate results. While this might seem straightforward, the nuances involved require both understanding and practice. This guide offers a step-by-step breakdown to ensure your initiation into prompt engineering is smooth and productive. Think of this as a best practice to get started with your prompt creation. I would highly suggest having a structured approach when getting started; with experience you will find new ways to create prompts.

Step-by-Step Instruction Guide

Step 1: Identify the Goal

Before anything, clarity on the desired outcome is essential.

Action: Clearly define the task. Are you looking for a concise answer, a deep dive on a topic, a creative exploration, or perhaps a specific dataset analysis?

Tip: Write down your objective in one sentence. For instance, "I want a brief overview of the history of AI."

Step 2: Start Simple

Your initial prompt should be clear but not overly complicated.

Action: Using your objective, craft a basic prompt. Example: "Give a brief history of AI."

Tip: Avoid ambiguity. A clear prompt leads to a clear response.

Step 3: Gauge the Initial Response

Before refining the prompt, observe the AI's first take.

Action: Input your simple prompt and analyze the response.

Tip: Take note of any unexpected outputs or deviations from your goal.

Step 4: Refine and Contextualize

Based on the AI's response, it's time to refine.

Action: Add specifics to your prompt or provide a context. Instead of the above example, try: "Provide a concise summary of AI's evolution from the 1950s to 2000s."

Tip: Context is crucial. The more specific the context, the narrower and more relevant the response usually is.

Step 5: Test for Variability

AI models can provide different outputs for the same prompt. It's beneficial to understand the range of potential responses.

Action: Input your refined prompt multiple times.

Tip: If you observe wildly varying outputs, consider making your prompt even more specific.

Step 6: Iterate and Adapt

Prompt engineering is an iterative process. Continual refinement is key.

Action: Based on the feedback from the previous step, further refine your prompt. For instance, if the AI missed out on some key developments in the history of AI, ask it explicitly to include them.

Tip: Iterative refinement is where your skills will grow the most. It's like a feedback loop between you and the AI.

Step 7: Utilize Advanced Parameters (If Available)

Many platforms offer parameters such as "temperature" or "max tokens" that affect AI responses.

Action: Adjust these parameters to see how they influence the results. A lower temperature, for example, makes the output more deterministic, while a higher value makes it more random.

Tip: Use these parameters to tailor the AI's response to your precise needs but always within the limits of ethical usage.

Step 8: Broaden Your Horizons

Once you've mastered a specific prompt, challenge yourself with different objectives.

Action: Switch topics, or change the nature of your request, moving from factual to creative or vice versa.

Tip: The more varied your prompts, the better your overall understanding and command of prompt engineering will be.

Step 9: Document and Learn

Keep track of your prompts and the AI's responses.

Action: Maintain a record of successful prompts and those that required significant refinement. Over time, this can become your personal playbook.

Tip: Observing patterns in AI behavior against certain prompts will allow you to predict its responses better, enhancing your prompt-engineering skills.

Step 10: Share and Collaborate

Prompt engineering is an evolving field. Collaboration and community engagement can provide valuable insights.

Action: Engage in forums, attend workshops, or collaborate with peers. Sharing experiences can offer fresh perspectives.

Tip: Different industries may use AI differently. Gleaning insights from varied sectors can be enriching.

Embarking on the journey of prompt engineering is akin to learning a new language. While you're essentially guiding an AI system in a particular direction, it's a dance of precision, patience, and iterative learning. Remember, there's no one-size-fits-all solution, and what works best often depends on the unique interplay between the specific AI model, the nature of the task, and the nuances of the prompt. As you dive deeper into this realm, you'll not only master the mechanics but also develop an intuitive sense of how to elicit the best from AI, making your endeavors both productive and insightful.

Testing and Evaluating Your Prompts

The art of prompt engineering doesn't stop at crafting a prompt; it extends to assessing its efficacy. A well-phrased prompt can be the difference between garnering a precise, actionable answer and receiving a vague or irrelevant one. Here's a guide on how to systematically test and evaluate your prompts for AI systems.

Understand the Importance

Testing isn't just a concluding step; it's an integral part of the prompt engineering process.

Action: Before diving into evaluations, recognize why it's crucial to ensure consistency, accuracy, and relevance in AI outputs.

Tip: Remember, a single prompt can yield different responses based on the AI's interpretation. Testing helps in narrowing down these variations.

Conduct Dry Runs

Start with initial evaluations before deploying prompts in real-world scenarios.

Action: Input your crafted prompt into the AI system multiple times. Analyze the range and consistency of answers.

Tip: Look for outliers or unexpected answers as they can provide insights into potential refinements.

Employ the A/B Testing Approach

This methodology involves comparing two versions to determine which performs better.

Action: Create two slightly different prompts aiming for the same outcome. Gauge which yields more accurate or comprehensive results.

Tip: This method can help in fine-tuning phrasing or context nuances.

Gather Peer Feedback

Sometimes, a fresh pair of eyes can offer invaluable insights.

Action: Share your prompts with colleagues or peers and have them evaluate the responses they receive.

Tip: Diverse perspectives can highlight unseen ambiguities or potential improvements.

Use Quantitative Metrics

While qualitative analysis is crucial, numeric metrics offer objective evaluations.

Action: Consider metrics such as response length, time taken for the AI to respond, or, if applicable, accuracy percentage.

Tip: Pairing qualitative insights with quantitative data provides a holistic view of a prompt's efficacy.

Consider Diverse Scenarios

Prompts might be interpreted differently based on scenarios or datasets.

Action: Test your prompt across varying scenarios or data samples to ensure it remains effective.

Tip: This ensures the prompt's robustness, especially critical for applications in dynamic environments.

Reiterate Based on Outcomes

Prompt engineering is an iterative process, and evaluations should feed into subsequent refinements.

Action: Based on testing outcomes, re-craft your prompt and test again.

Tip: Iterative refinement often leads to the most optimized prompts.

Understand Model Limitations

AI models, regardless of their sophistication, have limitations.

Action: If certain prompts consistently fail or yield suboptimal answers, it might be due to model constraints and not the prompt itself.

Tip: Familiarize yourself with the model's strengths and weaknesses. Some challenges might necessitate model retraining or the adoption of a different model.

Document Learnings

Maintaining a record of your evaluations can provide insights for future prompt crafting.

Action: Create a repository of tested prompts, their outcomes, feedback received, and refinements made.

Tip: This "knowledge bank" can be a quick reference for understanding what works and what doesn't in your specific use case.

Stay Updated

As AI models evolve and new techniques emerge, the landscape of what's possible with prompting can shift.

Action: Keep an eye on the latest research, attend workshops, or participate in online forums dedicated to prompt engineering.

Tip: Continual learning ensures you're leveraging the latest methodologies in your evaluations.

Testing and evaluating prompts is a meticulous yet rewarding endeavor. It's akin to polishing a lens: each refinement brings the world into sharper focus, ensuring AI models provide the clearest, most relevant insights. While the journey might involve trials and errors, each iteration brings you closer to mastering the nuances of prompt engineering. As AI continues its ascent in influencing diverse sectors, those adept at crafting, testing, and evaluating prompts will be at the forefront, harnessing the power of AI with precision and purpose.

Iterating and Refining Your Prompts

Prompt engineering, akin to any skill, requires continual refinement. While an initial prompt might seem perfect, there's always room for improvement to get the most accurate and insightful results from AI models. Here's an in-depth look into iterating and refining prompts within the realm of prompt engineering.

Understand the Need for Iteration

The dynamic nature of data, AI models, and objectives necessitate iterative refinement.

Action: Once you've created a prompt, treat it as a draft rather than a finished product.

Tip: By remaining open to adjustments, you ensure your prompts evolve in tandem with your project's needs.

Analyze the AI's Output

Every AI response offers a clue about the prompt's effectiveness.

Action: Scrutinize the output. Is it too general, too verbose, lacking detail, or deviating from the topic?

Tip: Highlight sections of the response that require refinement. These will guide your next prompt iteration.

Enhance Specificity

Often, vague outputs stem from generic prompts.

Action: Make your prompt more explicit. If you asked for "information on whales," try "the migratory patterns of blue whales."

Tip: Specificity narrows down AI's focus, yielding more targeted results.

Adjust the Phrasing

Sometimes, it's not about what you ask but how you ask.

Action: Experiment with different phrasings. Instead of "Describe Paris," try "Give a historical overview of Paris."

Tip: Different phrasings can elicit varied depth and perspectives from the AI.

Provide Context

AI models benefit from contextual data.

Action: Add relevant background information to your prompt. Instead of "effects of caffeine," specify "effects of caffeine on human sleep patterns."

Tip: Contextual cues guide the AI to generate responses aligned with your desired framework.

Test across Multiple Scenarios

A well-refined prompt works effectively across diverse datasets and scenarios.

Action: Apply the prompt in different use cases or subjects to assess its robustness.

Tip: This ensures versatility, especially vital for broad applications.

Seek External Feedback

A fresh perspective can offer novel insights.

Action: Share your prompt and the AI's response with colleagues or peers for their feedback.

Tip: Diverse opinions can highlight overlooked ambiguities or potential areas of enhancement.

Monitor Evolving Objectives

As projects progress, objectives might shift.

Action: Regularly revisit and assess whether your prompts align with your project's current goals.

Tip: An adaptive prompt strategy ensures you're always on track with your evolving objectives.

Leverage Advanced Features

Some platforms offer additional tools to fine-tune AI responses.

Action: Utilize features such as "temperature" adjustments or "max tokens" to refine outputs.

Tip: While the core prompt remains crucial, these features can fine-tune responses to your desired granularity.

Document and Review

Maintaining a log of your prompts and refinements is invaluable.

Action: Create a record of each prompt iteration, the AI's response, and the subsequent refinements made.

Tip: This documentation serves as a learning tool, providing insights into patterns of effective prompting.

Iteration and refinement in prompt engineering are not just strategies—they're necessities. As you delve deeper into the intricacies of AI interactions, you'll realize that the most effective prompts are often those that have undergone multiple cycles of refinement. It's a dynamic dance, with each step, twist, and turn enhancing the final outcome. And while this process demands patience and persistence, the results—a finely tuned, precise AI response—are well worth the effort. As AI continues to permeate various sectors, mastering the art of iterative prompt refinement stands as a cornerstone skill, ensuring you harness the full potential of AI models, aligning them seamlessly with human objectives and aspirations.

Prompt Engineering for Various Applications

Prompt engineering, at its core, is about optimizing AI behavior to align with specific requirements. Depending on the application, the nature and structure of prompts can vary considerably. This guide elucidates how prompt engineering can be tailored for various domains, ensuring effective and relevant AI interactions.

Customer Support Chatbots

Chatbots have revolutionized customer support by providing real-time assistance.

Action: Frame prompts to facilitate problem solving. Example: Instead of "What's the issue?" use "Please describe the issue you're facing with our product."

Tip: Direct, solution-oriented prompts yield actionable insights, expediting problem resolution.

Research and Academic Analysis

AI can assist researchers in sifting through vast datasets or summarizing complex studies.

Action: Craft prompts that ask for detailed analyses or concise summaries based on the need. Example: "Provide a summary highlighting the main findings of the study on XYZ."

Tip: Clear instructions ensure the AI's output aligns with academic rigor and relevance.

Creative Writing and Content Generation

Whether it's generating story ideas or enhancing narratives, AI can be a boon for writers.

Action: Guide the AI toward creativity with open-ended prompts. Example: "Craft a suspenseful intro for a story set in a haunted mansion."

Tip: Balancing specificity with freedom lets the AI's creativity shine while adhering to the desired theme.

Financial Analysis

In finance, prompt engineering can assist in extracting key metrics or forecasting trends.

Action: Make prompts data-specific. Example: "Analyze the quarterly earnings of company X and forecast the annual trend."

Tip: Precision is paramount; ensure prompts request the exact data points needed.

Medical Diagnostics Assistance

AI models can assist doctors by analyzing medical data.

Action: Frame prompts to seek specific medical insights. Example: "Compare the patient's symptoms with common flu indicators."

Tip: In medical applications, clarity and accuracy in prompts are crucial, given the stakes involved.

Language Translation

Translating languages is another domain where AI shines.

Action: Direct the AI clearly on the source and target languages. Example: "Translate the following English text to French: . . ."

Tip: Providing context can help. For a colloquial phrase, add "(informal)" to guide the AI's translation tone.

Entertainment and Media

From script suggestions to music composition, AI has a role in entertainment.

Action: Tailor prompts to the specific entertainment medium. Example for music: "Compose a cheerful tune suitable for a summer day."

Tip: Provide mood, genre, or other guiding factors to shape the AI's creative output.

Legal Applications

AI can assist in drafting legal documents or analyzing case laws.

Action: Make prompts explicit and detail-oriented. Example: "Draft a rental agreement for a residential property in California for a one-year lease."

Tip: Given the importance of accuracy in legal texts, always ensure clarity in prompts.

E-commerce and Retail

AI can assist in product descriptions, customer queries, or inventory management.

Action: Frame prompts based on the retail requirement. Example for product descriptions: "Write a detailed description for a men's leather jacket, highlighting its features."

Tip: In e-commerce, prompts should cater to both informativeness and appeal.

Urban Planning and Architecture

AI can assist in analyzing urban data or suggesting architectural designs.

Action: Guide the AI based on the architectural need. Example: "Suggest a sustainable design for a community park in a tropical city."

Tip: Providing context, such as location, climate, or cultural nuances, can refine the AI's suggestions.

As the horizons of AI expand, prompt engineering's role in shaping its interactions across domains becomes increasingly pronounced. Whether it's crafting a compelling narrative or forecasting stock market trends, the way we engage with AI through prompts determines the value and relevance of its outputs. By tailoring our approach to the unique demands of each application, we don't just optimize AI's responses—we also open the door to innovative solutions, insights, and possibilities. As AI continues to evolve, prompt engineering stands as a beacon, guiding our journey in harnessing AI's potential to its fullest across myriad applications.

Tips and Tricks for Advanced Prompt Engineering

As AI models grow increasingly complex and versatile, the intricacies of prompt engineering evolve alongside. Mastering advanced techniques becomes paramount for those looking to harness the full power of AI. Here's a deep dive into advanced strategies for prompt engineering, ensuring you stay ahead of the curve.

Multistep Prompts

Complex queries might require breaking down into multiple steps.

Action: Chain related prompts together to guide the AI through a series of thoughts. Example: First ask about the history of a topic, and then delve into its modern implications.

Tip: This approach helps in obtaining detailed, layered responses, especially for complex topics.

Utilize Contextual Tokens

Certain platforms allow the use of tokens to provide contextual clues.

Action: Embed tokens such as <informal> or <technical> to guide the AI's tone and style.

Tip: Such tokens can help obtain outputs that match the desired tone without explicitly stating it in the prompt.

Layered Refinement

When working with intricate topics, refine your prompt in stages.

Action: Begin with a broad prompt. Based on the AI's response, refine your query, making it more specific, and repeat.

Tip: This iterative method often yields richer insights than a single, direct prompt.

Exploit the "Nudging" Technique

Sometimes, slight nudges can guide the AI to desired outputs.

Action: If the AI is drifting off-topic, integrate phrases such as "in summary" or "to be precise" to refocus the model.

Tip: Subtle linguistic nudges can reorient the AI without a complete prompt overhaul.

Active Feedback Loops

In real-time applications, utilize user feedback to enhance prompts dynamically.

Action: Implement a feedback mechanism where users can rate or comment on AI outputs. Use this data for prompt adjustments.

Tip: Continuous feedback ensures the AI remains attuned to users' evolving needs.

Prompt Templates

For recurring tasks, design prompt templates.

Action: Create a basic structure for common queries, with placeholders to insert specific data.

Tip: Templates expedite prompt crafting, ensuring consistency and saving time.

Limit and Guide AI Exploration

AI models can sometimes generate diverse outputs. Direct their exploration for relevant results.

Action: Use features such as "temperature" for controlling randomness and "max tokens" for response length.

Tip: Setting boundaries helps in obtaining consistent, focused outputs while maintaining AI's generative prowess.

Explore AI's Metacognition

Advanced models can reflect on their own knowledge.

Action: Craft prompts that ask the AI about its limitations or areas of expertise. Example: "What do you know about quantum computing?"

Tip: Such meta-prompts can provide insights into the AI's capabilities, guiding subsequent queries.

Parallel Testing

When unsure about the best phrasing, run multiple prompts in parallel.

Action: Craft different versions of the same prompt and assess which yields the best results.

Tip: Parallel testing offers insights into optimal phrasing and can be automated for efficiency.

Stay Abreast with Evolving Models

AI research is dynamic. Regularly update your prompt-engineering skills.

Action: Engage with communities, attend workshops, or follow the latest AI research.

Tip: Understanding the underlying model changes helps in crafting more effective prompts.

The world of advanced prompt engineering is vast and continually evolving. As we stand at the intersection of linguistics, technology, and human intuition, the strategies we deploy play a pivotal role in determining the quality of AI interactions. By embracing advanced techniques, we ensure that AI not only understands our queries but responds in ways that are insightful, relevant, and transformative. As the next wave of AI innovations beckons, armed with these advanced strategies, we are poised to navigate the landscape with finesse, turning AI's potential into palpable solutions that resonate with real-world challenges and aspirations.

Advanced Techniques: Machine Learning for Prompt Optimization

In the realm of prompt engineering, machine learning (ML) has emerged as a powerful ally, pushing the boundaries of what's possible. By integrating ML techniques, we can optimize prompts dynamically, making interactions with AI models more precise and insightful. Let's delve into how ML can elevate the art and science of prompt engineering.

Why ML?

Traditional prompt engineering is a blend of intuition and trial-and-error. ML offers systematic refinement.

Action: Instead of manual adjustments, employ ML algorithms to analyze AI responses and refine prompts automatically.

Tip: The marriage of prompt engineering and ML caters to scalability and automation, especially vital for large-scale applications.

Supervised Learning for Prompt Refinement

Harness labeled datasets to train models for prompt optimization.

Action: Using historical data of prompts and their AI responses, train an ML model to suggest prompt refinements.

Tip: Over time, this model learns from successes and failures, enhancing the quality of prompts.

Reinforcement Learning (RL) in Action

RL offers a dynamic approach where models learn by interacting with the environment.

Action: Implement an RL agent to craft and refine prompts. Reward it based on the quality of AI responses.

Tip: The agent learns to adjust prompts to maximize rewards, leading to continual improvement.

Hyperparameter Tuning for Prompts

Just as ML models have hyperparameters, prompts have nuances that can be tuned.

Action: Adjust elements such as prompt length, specificity, or context. Use ML to find the optimal combination.

Tip: This ensures that each prompt is tailored for maximum efficacy.

Clustering for Response Analysis

Categorize AI responses to understand common patterns and outliers.

Action: Use clustering algorithms such as K-means to group similar AI outputs. Analyze each cluster to refine prompts.

Tip: This technique highlights consistent issues or successes in AI responses, guiding prompt adjustments.

Natural Language Processing (NLP) for Insight Extraction

Leverage NLP to dissect AI responses and extract key insights.

Action: Implement NLP tools to analyze the semantics and sentiment of AI outputs, gauging their relevance and accuracy.

Tip: NLP can pinpoint areas where the AI might be misunderstanding or oversimplifying, guiding prompt refinement.

Feedback Loops with Active Learning

Integrate user feedback into the ML model for real-time refinement.

Action: Allow users to rate or comment on AI responses. Feed this data into the ML model to adjust prompts dynamically.

Tip: Active learning ensures that the system adapts to user preferences and needs over time.

Generative Models for Prompt Creation

Why limit AI to responding? Let it aid in prompt crafting too!

Action: Train generative models to suggest potential prompts based on desired outcomes.

Tip: This can be especially useful for beginners in prompt engineering, offering a starting point for their queries.

Transfer Learning for Rapid Adaptation

Benefit from preexisting ML models and knowledge.

Action: Use transfer learning to apply knowledge from one prompt engineering task to another.

Tip: This technique expedites the learning process, especially when moving between similar tasks or domains.

Periodic Model Evaluation and Retraining

The landscape of AI and data is dynamic. Ensure your ML models stay updated.

Action: Regularly evaluate the performance of your ML models for prompt optimization. Retrain them with fresh data as needed.

Tip: Consistent evaluations ensure that the model's suggestions remain relevant and effective.

ML's infusion into prompt engineering is a testament to the evolving synergy between human expertise and computational prowess. By automating the intricate process of prompt refinement, we unlock a realm where AI interactions are more seamless, insightful, and aligned with our objectives.

The blend of prompt engineering and ML doesn't just enhance the quality of AI outputs—it paves the way for a future where our dialogues with machines are increasingly nuanced, meaningful, and transformative. As we embark on this exciting journey, the fusion of ML techniques ensures we navigate the challenges with precision, turning the vast potential of AI into tangible solutions that resonate across applications and industries.

Ethical Considerations in Prompt Engineering

Table of Contents

Understanding Bias in AI and Prompts

The concept of AI has often been touted as a potential panacea for various challenges. However, as AI technologies, particularly language models, become more embedded in our daily lives, the subject of biases in these models and their impact on society has become a significant concern. This discussion delves into the nuances of bias within AI and how it manifests in the world of prompt engineering.

At its core, a language model learns from vast quantities of data. This data, often sourced from the Internet, reflects human-written content and inherently carries the biases of its authors. Thus, when the model is trained, it imbibes these biases. For instance, if a particular gender, race, or group is predominantly portrayed in a specific manner in the training data, the model will likely mirror these biases in its outputs.

Prompt engineering is a pivotal area within AI where biases can either be mitigated or exacerbated. The way a prompt is framed can heavily influence the response generated by an AI model. By understanding the potential pitfalls and being aware of existing biases, prompt engineers can attempt to craft unbiased and objective prompts.

However, this is easier said than done. Here are a few challenges of AI in a very generalized way.

Data Imbalance

The training data for many language models often comes from vast online repositories, which might not represent minority views or might be skewed toward popular opinions. This lack of balanced representation can lead to an AI model that inadvertently favors certain perspectives over others.

Historical Biases

Some biases are deeply rooted in historical contexts. Even if contemporary data might be more balanced, the lingering effects of past biases can still affect AI responses.

Subtle and Unintentional Biases

Not all biases are overt. Sometimes, they can be subtle and may not even be apparent to the prompt engineer. For instance, seemingly neutral terms or phrases can carry different connotations in various cultural contexts.

Feedback Loops

If an AI's biased output is not checked and corrected and it continually receives biased feedback, it can end up reinforcing the same biases. This creates a vicious cycle where the AI becomes more ingrained in its prejudiced views.

The repercussions of biased AI are manifold. It can lead to unfair treatment, reinforce stereotypes, and even cause harm in certain situations. For instance, in a job recruitment scenario, if the AI has gender or racial biases, it can unfairly favor or disadvantage certain candidates. This not only robs deserving candidates of opportunities but also deprives organizations of potential talent.

Given the potential harm, addressing bias in AI, especially in prompt engineering, is of paramount importance. Here are a few steps that can be taken.

Awareness and Training AI practitioners, especially prompt engineers, should be educated about the existence and implications of biases. Being aware is the first step toward mitigation.

Diverse Training Data Efforts should be made to curate a diverse and balanced dataset. This can reduce the chances of the AI model inheriting skewed views.

Regular Audit Even after training, AI models should be regularly audited for biases. Any biased behavior should be flagged, analyzed, and rectified.

Feedback Mechanisms Encourage users to report biased outputs. This feedback can be invaluable in identifying and rectifying biases.

Collaborative Efforts Addressing biases in AI is not the sole responsibility of a single entity or organization. It requires a collective effort, involving researchers, practitioners, policymakers, and users.

In conclusion, biases in AI, and by extension in prompts, are a reflection of our society. While it's challenging to create a completely unbiased AI, being cognizant of the biases and making concerted efforts to minimize them can ensure that AI technologies are fair, objective, and truly beneficial for all.

Strategies for Reducing Bias in Your Prompts

The increasing ubiquity of AI models in our lives, combined with their inherent potential for biases, has made it vital for prompt engineers to design systems that are not only intelligent but also equitable. While complete eradication of biases is an aspirational goal, there are clear strategies that can substantially reduce their occurrence and impact. This discourse explores methods that can be employed to ensure that the prompts used in AI interactions remain as unbiased as possible.

Comprehensive Training

Before diving into practical strategies, it's paramount that prompt engineers are educated on potential biases. Comprehensive training, which includes understanding sociocultural contexts, can empower engineers to identify subtle biases they might otherwise overlook.

Diverse Input Review

Once a prompt is designed, have it reviewed by a diverse group. Different perspectives can shed light on unintentional biases that the original designer might not be aware of. Encouraging diversity in the design and review process ensures a wider array of challenges and considerations are addressed.

Use Neutral Language

Avoid using language that implies gender, age, ethnicity, or any other potential point of bias unless it's strictly relevant to the task. For instance, instead of using "he" or "she," "they" can be used as a gender-neutral pronoun.

Fact-Based Design

Ensure that the prompts are designed based on facts and data rather than assumptions or stereotypes. For example, avoiding stereotypes about particular groups of people when asking AI to generate stories or examples can reduce biased outputs.

Dynamic Adaptability

Design prompts that can adapt to users' feedback or concerns about biases. This not only provides a mechanism to correct the model but also builds trust with users, as they can see their feedback being taken into account.

Use of Anti-bias Tools

There are various tools available that can detect and highlight biased terms or phrases in text. Incorporating such tools in the prompt design process can act as an additional layer of verification.

Thorough Testing

Before deploying, test the prompts in diverse scenarios and with varied user groups. This real-world testing can shed light on biases that might not be evident in controlled environments.

Iterative Refinement

Understand that no prompt is perfect from the get-go. Continuously gather feedback, learn from shortcomings, and refine the prompts. This iterative process, combined with the feedback loop from users, can significantly improve the quality and fairness of prompts over time.

Transparency and Openness

Be transparent about the potential biases and the steps taken to mitigate them. Providing users with information on how the AI system operates and the kind of data it was trained on can set the right expectations and foster trust.

Collaboration with Ethicists

Engage with ethicists and sociologists who specialize in technology and its societal impacts. Their insights can provide valuable perspectives on the ethical implications of prompt decisions and guide engineers toward more equitable solutions.

Community Engagement

Engaging with the wider AI community, including researchers, practitioners, and users, can provide a plethora of insights. Open forums, workshops, and discussions can be platforms for sharing experiences, challenges, and best practices in reducing biases.

Setting Ethical Standards

Organizations should set clear ethical standards for prompt engineering. These standards, which should be continuously updated to reflect evolving societal norms and understandings, can act as a guidepost for engineers.

Documented Feedback Mechanism

Having a clearly documented process where users can provide feedback about biased or inappropriate outputs can help in iterative refinement. It also ensures that users feel valued and heard.

The quest to reduce biases in prompts is both challenging and continuous. As societies evolve and as AI technologies become more sophisticated, the strategies to address biases need to adapt as well. The key lies in maintaining a balance between leveraging AI's capabilities while ensuring its equitable application. Through conscious efforts, continuous learning, and collaboration, the AI community can stride toward more unbiased and just systems.

Ethical Guidelines for Prompt Design

In the rapidly evolving landscape of AI and ML, ethical considerations have emerged as a critical focal point. Given the significant impact of prompts on AI behavior, there's a pressing need for ethical guidelines in their design. Here's an in-depth exploration of ethical standards that should be central to prompt engineering.

Prioritize Fairness and Avoid Discrimination

Relevance: AI systems are inherently neutral, but their outputs are shaped by the data they're trained on and the prompts they receive. Without careful consideration, these outputs can unintentionally propagate biases present in the training data.

Action: Strive for fairness in prompt design by avoiding language or instructions that can lead to discriminatory outputs. Prioritize inclusivity and ensure the AI system doesn't favor any group over another.

Maintain Transparency

Relevance: The black box nature of many AI models can make their decisions opaque, causing trust issues among end users.

Action: Design prompts that produce transparent and interpretable results. Users should be able to understand the basis of the AI's response.

Protect User Privacy

Relevance: As AI systems interact with users, there's potential for these systems to access sensitive or personal information.

Action: Design prompts that do not solicit private information. If data collection is necessary, ensure explicit consent and explain the purpose of data gathering.

Safeguard against Harmful Outputs

Relevance: In some cases, seemingly innocent prompts can lead AI to generate harmful or inappropriate content.

Action: Implement safeguards to prevent or flag potential harmful outputs. Testing the AI system extensively can also help identify and rectify such issues.

Ensure Accountability

Relevance: When AI systems make mistakes or when their outputs have unintended consequences, there should be a clear line of accountability.

Action: Clearly document the design process, decisions made, and the rationale behind prompts. This documentation can help trace back and rectify any issues.

Promote User Autonomy

Relevance: AI is a tool, and its purpose is to assist, not dominate or manipulate.

Action: Design prompts that empower users, providing them with information or options, rather than taking control away from them.

Ensure Cultural Sensitivity

Relevance: With AI being globally accessible, there's a chance that it might inadvertently offend cultural or social norms.

Action: Understand the diverse cultural contexts in which the AI will operate and design prompts accordingly. This might involve localizing prompts for different regions or cultures.

Avoid Reinforcing Stereotypes

Relevance: There's a risk that AI, based on its training data, might perpetuate stereotypes.

Action: Craft prompts that challenge or neutralize these stereotypes rather than reinforce them. Ensure the AI system is inclusive in its responses.

Practice Continual Learning and Adaptation

Relevance: The field of AI is dynamic, with societal norms and technological capabilities continuously evolving.

Action: Regularly update and refine prompts in line with new knowledge, societal feedback, and technological advancements.

Engage with a Diverse Group

Relevance: A homogenous group of designers might inadvertently introduce biases into the system.

Action: Engage with a diverse set of designers, users, ethicists, and other stakeholders when crafting prompts. Different perspectives can identify and rectify potential pitfalls.

Implement Ethical Review Processes

Relevance: Just as many research projects undergo ethical reviews, AI systems, given their potential impact, should also be scrutinized.

Action: Establish a process for ethical review of prompts, particularly in sensitive applications. This review can act as a final checkpoint to ensure ethical standards are met.

Educate and Empower Users

Relevance: Users should be aware of the capabilities and limitations of the AI system.

Action: Design prompts that educate users about how the system works, its potential biases, and how they can interact most effectively with it.

Commit to Long-Term Responsibility

Relevance: The responsibility of prompt designers doesn't end once the system is deployed.

Action: Continuously monitor the AI system's performance, gather user feedback, and make necessary adjustments to the prompts to address any emerging ethical concerns.

In conclusion, ethical considerations in prompt engineering are not just a matter of compliance or good PR; they're fundamental to building AI systems that are beneficial, fair, and trusted by users. By adhering to these guidelines, prompt engineers can pave the way for AI's responsible and beneficial integration into society.

Prompts and Privacy Considerations

The advent of AI has propelled an interactive ecosystem, enabling machines to simulate humanlike cognition and responses. Within this milieu, prompts have emerged as pivotal tools that help elicit specific responses from AI systems. However, the dynamic interface prompts also introduce privacy considerations that are crucial to acknowledge and address.

Collection of Personal Information

Prompts can often steer users to provide personal or sensitive information, either intentionally or inadvertently. While some data might enhance the user experience, it's crucial to determine the necessity of such data. If collecting personal data is essential, users should be thoroughly informed, and their explicit consent should be secured.

Storage and Security

The data solicited via prompts is usually stored for processing and potential future references. The sanctity of this stored data is paramount. Leveraging robust encryption techniques and secure storage solutions can thwart potential breaches. Users should always be apprised of where and how long their data will be kept.

Data Anonymization

Even seemingly innocuous data can, in aggregation, be used to trace back to individual users. Data anonymization helps strip away identifiable features, ensuring user identities remain concealed. Transparency about this process is essential to maintain user trust.

Data Sharing and Third-Party Access

Shared data can sometimes find its way to third parties, be it for analytics or other operational requirements. Clear articulation of such processes in prompt design can help users make informed decisions about their data.

Children's Data

The sensitivity of children's data amplifies the need for stringent controls and regulations. If a system is designed for or can be accessed by children, regulatory compliance and cautionary measures become even more essential.

Contextual Awareness

AI's ability to comprehend the context can be instrumental in preventing inadvertent data collection. For instance, in sensitive scenarios, the system can be trained to refrain from soliciting private information.

Consent Revocation

Empowering users to revoke data consent embodies an ethical design practice. Users should have a seamless experience in both granting and withdrawing consent, ensuring they remain in control of their data.

Legal and Regulatory Compliance

With global users comes the responsibility of adhering to a mosaic of privacy laws, each tailored to different regions. Familiarity and compliance with these laws ensure that prompt design respects user rights across borders.

Transparency and User Education

Beyond collecting data, it's ethical and beneficial to enlighten users about the ramifications of their data sharing. An informed user can make better decisions about their interactions with AI.

Ethical Boundaries

While data can be potent, establishing clear ethical boundaries ensures that data collection remains purposeful and doesn't verge into invasive territories.

Audit Trails

As with any system, having a trail of activities aids in troubleshooting and accountability. Maintaining a comprehensive record of all prompt interactions is not just beneficial for technical reasons but also for potential audits and reviews.

Feedback Loops

A system's true evaluation comes from its users. Their feedback can spotlight overlooked privacy concerns. An open channel for such feedback can thus be instrumental in iterative improvements.

Regular Updates and Reviews

The rapid evolution of technology and regulations necessitates frequent updates to prompts and underlying systems. Keeping abreast of changes and ensuring prompt designs reflect the most recent guidelines is key.

In essence, while prompts serve as gateways to richer AI-human interactions, they also shoulder the responsibility of safeguarding user privacy. Striking this balance calls for a blend of ethical considerations, technical prowess, and user-centric designs.

Future Ethical Challenges in Prompt Engineering

Prompt engineering, at its core, represents a liaison between human input and machine response. As technology continues its inexorable march forward, this symbiotic relationship grows in complexity, bringing forth ethical challenges. While the present already introduces nuances for us to contemplate, the future of prompt engineering will inevitably surface more profound ethical dilemmas.

Sophistication of AI Responses

As AI becomes more sophisticated, it will generate outputs that may be indistinguishable from human-crafted content. This could lead to situations where fabricated information appears genuine, potentially misleading audiences and spreading misinformation.

Hyper-personalization

As AI systems become adept at tailoring responses based on user data, there's a potential risk of creating echo chambers. Users may only receive information aligned with their current beliefs, stifling diverse viewpoints and encouraging confirmation bias.

Autonomy of AI Decisions

Future AI systems might have increased autonomy in decision-making. With sophisticated prompts, the line distinguishing a suggestion from a command might blur, creating ethical challenges when AI makes decisions that humans would deem inappropriate or harmful.

Manipulative Prompts

As we gain a deeper understanding of human behavior and psychology, there's a risk that prompts could be engineered to manipulate users subtly. This can be particularly concerning in areas such as advertising, political campaigns, or any sphere aiming to influence public opinion.

Inclusivity and Representation

As language models diversify to represent various cultures, languages, and demographics, ensuring that they don't perpetuate stereotypes or biases becomes a critical challenge. Designing prompts that are inclusive and culturally sensitive will be paramount.

AI as a Social Actor

There's growing evidence that humans anthropomorphize AI, treating them as social entities. This relationship could become problematic if people begin to develop unhealthy attachments or dependencies, based on the way prompts are engineered.

Ethics of Emotional AI

Predicted advancements might allow AI to better discern human emotions and respond accordingly. The ethical implications are vast, from potential therapeutic applications to concerns about manipulating human emotions for commercial or other gains.

Privacy Erosion

As AI becomes more intertwined in daily life, prompts might solicit more personal and intimate details. The future might pose challenges in discerning what is truly beneficial for user experience versus what infringes upon privacy.

Economic and Employment Implications

AI, guided by advanced prompts, might outperform humans in areas such as content creation, journalism, or design. The ethical challenges here involve the potential displacement of jobs and ensuring that economic benefits are distributed equitably.

Deepfake and Reality Distortion

With AI capable of generating highly realistic content, prompts can be used to create deepfakes or simulate real-world scenarios. The ethical ramifications of blurring the lines between reality and fiction are profound, especially in contexts like news dissemination or evidence in legal proceedings.

Transparency and Accountability

As AI models grow in complexity, so does their opacity. Ensuring transparency in how prompts influence AI, and holding creators accountable for the outputs, will be crucial.

Regulation and Censorship

With AI's pervasive influence, governments and organizations might seek to regulate or control prompts and outputs. This poses ethical challenges around freedom of expression, censorship, and the potential misuse of AI for propaganda.

Moral and Ethical Dilemmas

AI might eventually be used to make moral or ethical decisions, guided by human-designed prompts. Ensuring that these decisions align with broader human values without imposing a singular moral framework will be challenging.

In the horizon of prompt engineering, while technological advancements promise a myriad of opportunities, they also introduce complex ethical mazes to navigate. Being proactive in recognizing, discussing, and addressing these challenges is not just the responsibility of developers or businesses but of society as a whole. It calls for multidisciplinary collaboration, involving ethicists, sociologists, technologists, and users, to ensure that the AI of the future remains both revolutionary and responsible.

Advanced Techniques: Automated Bias Detection

Prompt engineering, the art of fine-tuning input to guide AI outputs, is pivotal in determining how ML models interact with users. Yet, as these systems become integral to our daily lives, ethical considerations have arisen, with bias detection being one of the most pressing concerns. Automated bias detection represents a promising frontier to tackle the pervasive issue of bias in AI. Here's an exploration of its facets.

Origins of Bias in AI

Before delving into automated detection, it's vital to understand the sources of AI bias. ML models are reflections of the data they're trained on. If training data contains biases—intentional or unintentional—the model will likely replicate and sometimes amplify them. With language models, bias can emerge from historical texts, popular culture, or any medium that captures societal values and prejudices.

Automated Bias Detection

At its core, automated bias detection leverages algorithms to identify, quantify, and sometimes rectify bias within AI models. By comparing model outputs across a diverse range of prompts and inputs, these algorithms can flag potential instances of bias, categorize them, and provide metrics on the model's fairness.

Benefits

Scalability: Manual bias audits are time consuming and limited in scope. Automated techniques can assess vast model architectures and huge datasets more rapidly.

Objectivity: Algorithms don't have personal biases, ensuring that detection is based on defined metrics and not personal perceptions.

Continuous monitoring: Automation allows for ongoing bias monitoring, ensuring that models remain fair as they evolve.

Techniques

Adversarial testing uses purposely crafted prompts to challenge the model and elicit biased responses. By consistently probing the model, one can identify its weak points.

Fairness metrics: These quantify disparities in model performance or outcomes across different groups, highlighting potential areas of concern.

Transfer learning and fine-tuning: Leveraging pretrained models that have undergone bias mitigation can serve as a starting point. Subsequent fine-tuning can further align the model with desired fairness criteria.

Challenges

Defining bias: The concept of bias is multifaceted and varies across cultures and societies. Algorithms need a clear definition to operate, which can be challenging.

Overcorrection: If not carefully managed, automated detection could lead to overcorrection, where attempts to remove bias lead to other forms of it or diminish the model's utility.

Transparency: The algorithms used for bias detection should be transparent. Black box solutions can make it hard to understand and trust the bias detection process.

Ethical Considerations

Privacy: Automated bias detection might require analyzing large amounts of user data, raising privacy concerns.

Accountability: Who is responsible if automated bias detection fails or introduces new biases? Ensuring accountability is vital.

Diversity: The teams developing these automated solutions should be diverse, ensuring a broad perspective and reducing unintentional oversights.

Applications in Prompt Engineering

By integrating bias detection during the prompt creation phase, engineers can receive immediate feedback, refining prompts iteratively. As users interact with models, automated bias detection can monitor outputs in real time, ensuring that user-facing applications maintain fairness standards.

Case Studies: Search Engines

Search engines automated bias detection can help in ensuring that search engine outputs are fair and don't favor any particular group or ideology. For platforms such as streaming services or news apps, bias detection ensures that recommended content isn't skewed due to underlying biases in algorithms.

The Road Ahead

While automated bias detection is promising, it is not a panacea. It should be a part of a broader tool kit, accompanied by manual reviews and user feedback. As AI models become more sophisticated, so should our methods of ensuring their fairness.

Automated bias detection represents a significant stride toward fairer, more ethical AI systems. By integrating these techniques into prompt engineering, we can guide AI models to produce outputs that respect the vast tapestry of human diversity and experience. The ethical mandate here is clear as creators and users of AI: continuous vigilance against bias isn't just beneficial—it's imperative.

Application-Specific Prompt Engineering

Table of Contents

Prompts for Creative Writing

Prompt engineering has revolutionized the field of AI, offering a structured way to guide AI models in producing desired outputs. In the field of creative writing, this capability offers a blend of human imagination and AI's vast knowledge base. Let's delve deeper into the fascinating interplay between prompts and creative writing.

Bridging Human Imagination with AI

Creative writing, inherently, is a reflection of human emotions, experiences, and narratives. Using prompts, we can guide AI to produce content that resonates with human sentiments. For instance, providing an

AI with a prompt such as "Write a story set in a post-apocalyptic world where humans and robots coexist" can produce a myriad of storylines infused with humanlike emotions and AI-driven creativity.

Genre-Specific Prompts

By specifying genres within the prompt, writers can utilize AI to generate content tailored to specific themes. A prompt such as "Write a horror tale about a haunted mansion" directs the AI toward horror-specific vocabulary and constructs, aiding authors in drafting spine-chilling narratives.

Inspiration and Idea Generation

For writers facing the dreaded writer's block, prompts can act as catalysts. An open-ended prompt such as "Describe a world where the sky is green" can generate imaginative settings and plots, serving as a starting point for authors to expand on.

Character Development

Prompts can also aid in character creation. Inputting a prompt such as "Sketch a character who can communicate with animals" can yield intricate character backgrounds, motivations, and story arcs, enhancing storytelling depth.

Dynamic Interaction

Prompting can be iterative. A writer can start with a general prompt, take the AI-generated content, refine it, and then use it as a new prompt to guide the AI into more detailed or different directions, creating a dynamic interaction between the writer and the AI.

Multimodal Applications

Beyond mere texts, prompts can also incorporate visual cues. For instance, a writer can use an image of a medieval castle and prompt the AI to "write a romantic tale based on this setting," marrying visual inspiration with textual creativity.

Ethical Considerations

It's essential to ensure that creative content generated is free from inadvertent biases or potentially harmful narratives. Crafting well-defined prompts becomes crucial to avoid misleading or inappropriate content.

Challenges and Limitations

While AI can enhance creative writing, it's vital to recognize its limitations. The outputs are based on training data, and the AI lacks genuine emotions or life experiences. Thus, the human touch remains irreplaceable. Over-reliance on AI can also lead to homogenized content, losing the unique voice of the individual author.

Future Potential

As AI models evolve, so will their ability to interpret and generate creative content. The synergy between human writers and AI tools will strengthen, offering writers enhanced tools for storytelling, character development, and world building.

Prompt engineering stands at the intersection of technology and creativity in the domain of creative writing. By guiding AI through meticulously crafted prompts, writers can harness the power of AI to expand their creative horizons, produce diverse narratives, and reshape the boundaries of storytelling.

In the rapidly evolving landscape of AI, prompt engineering for creative writing represents a confluence of human artistry and machine precision. As writers embrace this synergy, the future holds boundless possibilities for enriched narratives and broader storytelling horizons.

Prompts for Business Applications

In the contemporary business landscape, AI, spearheaded by sophisticated language models, has carved out a pivotal role. Prompt engineering, a subdiscipline, has further fine-tuned the capabilities of these models to cater to specific business needs. Let's delve into the utility of prompts in the realm of business applications.

Data Analysis and Insights

Modern businesses are inundated with data. AI models, when fed with prompts such as "Analyze sales data from Q1 and provide key trends," can sift through vast datasets, offering insights that might be missed by human analysts. Such prompts ensure targeted, relevant, and timely data interpretation, a vital asset for businesses aiming to maintain a competitive edge.

Customer Support and Interaction

Chatbots and virtual assistants are becoming ubiquitous in customer support roles. Prompts such as "Assist the user in troubleshooting printer issues" guide AI in offering step-by-step solutions, ensuring customer queries are resolved efficiently and effectively.

Market Research and Consumer Insights

Understanding market dynamics is crucial for business success. By using prompts such as "Extract consumer sentiments about our new product from online reviews," businesses can leverage AI to gauge consumer reactions, preferences, and pain points.

Content Generation

In the digital age, content is king. Companies can utilize prompts such as "Draft a blog post about the environmental benefits of our product" to generate relevant, coherent, and engaging content, catering to their target demographic.

Financial Forecasting

Financial acumen drives business decisions. With prompts such as "Predict stock market trends for the next month based on historical data," AI models can assist in financial forecasting, enabling businesses to strategize and plan more effectively.

Personalization in Marketing

Consumers increasingly desire personalized experiences. By employing prompts such as "Suggest product recommendations for a user who enjoys hiking," businesses can tailor their marketing efforts to individual preferences, enhancing user engagement and boosting sales.

Human Resources and Recruitment

The hiring process is resource-intensive. Using prompts such as "Screen résumés for candidates with more than five years of coding experience," AI can streamline the recruitment process, ensuring only the most relevant candidates are shortlisted.

Ethical Implications

It's paramount that businesses employ AI responsibly. Using biased or inappropriate prompts can lead to skewed results, damaging a company's reputation. Carefully engineered prompts ensure ethical and unbiased AI utilization, aligning with a business's values and societal norms.

Challenges

While prompts can guide AI toward desired outputs, the nuances of the business environment, governed by human emotions, cultural factors, and ever-evolving trends, present challenges. Over-reliance on AI, without human oversight, can lead to lapses, misinterpretations, and missed opportunities. Crafting prompts that encapsulate the multifaceted business ecosystem is both an art and a science.

Future Trajectories

As AI models become more sophisticated, the scope of prompt engineering in business applications will expand. We can envisage a future where prompts guide AI in complex negotiations, strategy formulations, and even in innovation. The symbiosis between human business acumen and AI's computational prowess, steered by meticulously crafted prompts, will redefine the business landscape.

Prompt engineering, while a technical discipline, has profound implications in the business domain. As companies strive to stay agile, innovative, and customer-centric in a dynamic global market, the ability to harness AI's capabilities through targeted prompts will emerge as a game changer. From enhancing operational efficiencies to driving innovation, prompts stand at the nexus of technology and business strategy, heralding a new era of data-driven decision-making and personalized consumer experiences.

Prompts for Educational Uses

In an age where education is increasingly blended with technology, the role of artificial intelligence has become ever more pronounced. Within this synthesis of AI and learning, prompt engineering emerges as a transformative tool, facilitating educational interventions tailored to diverse learning needs. The meticulous crafting of prompts has the potential to reshape how educators and students interact with AI-driven educational tools.

Personalized Learning Paths

Traditional classroom settings often follow a one-size-fits-all approach. However, with AI, prompts such as "Design a learning path for a student struggling with algebra" can guide the system to curate customized lesson plans, addressing individual student weaknesses and optimizing learning trajectories.

Tutoring and Homework Assistance

AI-driven platforms can act as virtual tutors. Prompts such as "Explain the Pythagorean theorem in simple terms" can guide the AI to deliver concise and understandable explanations, catering to students who might find certain concepts challenging.

Language Learning

For students learning new languages, prompts such as "Provide conversational practice for a beginner-level French learner" can facilitate interactive and immersive language experiences, enhancing fluency and comprehension.

Special Education Needs

AI, guided by prompts, can be a boon for special education. For instance, "Adapt this story for a student with dyslexia" can lead to content that's more accessible, with features such as simplified language or integrated audio descriptions.

Interactive Simulations

For subjects such as physics or chemistry, prompts such as "Simulate a chemical reaction between hydrogen and oxygen" can initiate interactive visualizations, aiding in conceptual clarity and offering hands-on virtual lab experiences.

Assessment and Feedback

AI can assist educators in grading and providing feedback. Prompts such as "Evaluate this essay for coherence and argument strength" can provide detailed feedback, highlighting areas of improvement and significantly reducing the workload of educators.

Augmenting Classroom Discussions

AI can be used to spark classroom discussions. A prompt such as "Suggest a debate topic on ethical implications of genetic engineering" can guide AI to come up with engaging and relevant discussion themes, fostering critical thinking among students.

Historical and Cultural Contexts

For history or literature lessons, AI can provide rich contextual backdrops. A prompt such as "Describe the socioeconomic landscape of 18th century France" can yield detailed expositions, making lessons more immersive.

Challenges

Despite the myriad benefits, there are inherent challenges. One primary concern is ensuring that AI-generated content aligns with educational standards and curricula. Over-reliance on AI without educator intervention might lead to gaps in learning. Moreover, crafting prompts that cater to diverse learning styles and cultural contexts requires deep pedagogical insights.

Ethical Implications

There's a fine line between aiding and replacing human intervention in education. While prompts can guide AI to provide assistance, the human touch— the essence of teaching—and mentoring mustn't be overshadowed. Additionally, data privacy, especially concerning minors, is paramount. Educators must ensure that AI tools, guided by prompts, adhere to strict privacy norms, protecting student data.

Future Directions

As AI becomes more sophisticated, the role of prompt engineering in education will expand. Future classrooms might see AI assistants, guided by real-time prompts, providing on-the-spot assistance, enriching classroom interactions. Moreover, with virtual reality and augmented reality becoming mainstream, prompts can guide AI to create immersive educational experiences, from historical reenactments to intricate scientific simulations.

The confluence of prompt engineering and education symbolizes the next frontier in pedagogical evolution. While the possibilities are boundless, a measured and ethical approach is crucial. As educators and technologists collaborate, crafting prompts that encapsulate educational objectives and learner needs, a new era of enriched, personalized, and interactive learning beckons. The synergy of human expertise and AI assistance, mediated by well-engineered prompts, has the potential to redefine educational paradigms, making learning more accessible, engaging, and effective.

Prompts for Coding and Software Development

In the domain of software engineering and coding, artificial intelligence is not merely a complementary tool but a transformative force. Prompt engineering, a nuanced way to guide AI's actions, plays an integral role in harnessing this force for the intricate processes involved in coding, debugging, and system design. Through tailored prompts, developers can delegate, automate, and enhance many of their tasks, making the software development life cycle more efficient and robust.

Code Generation

With advanced models such as OpenAI's Codex, developers can use prompts to auto-generate code snippets. A prompt such as "Generate a Python function to sort a list of numbers" would return a functional piece of code, reducing manual coding effort.

Debugging Assistance

Bugs are an inevitable part of coding. Instead of manually sifting through lines of code, developers can use prompts such as "Identify syntax errors in the following JavaScript function" to pinpoint and rectify errors.

Code Optimization

Beyond just writing functional code, optimization is crucial for performance. Prompts such as "Optimize the following SQL query for better performance" can guide AI tools to suggest or rewrite sections of code for efficiency.

System Design and Architecture

While initially thought of as a strictly human-driven task, with the right prompts, AI can suggest system architectures. For instance, "Design a microservices architecture for an e-commerce platform" could yield a basic design schema, serving as a starting point for architects.

Integration and API Usage

Integration with other systems or using third-party APIs can be complex. Prompts such as "Generate a Python code to connect to XYZ API and fetch data" can expedite this process, ensuring seamless integration.

Documentation and Commenting

Often overlooked but vital, documentation can be facilitated using AI. A prompt such as "Provide a documentation template for the following Java class" can aid developers in creating comprehensive and standardized documentation.

Code Review and Quality Assurance

Using prompts, AI can be directed to review code for best practices. "Review the following C++ code for adherence to standard conventions" can result in feedback on where the code deviates from set standards.

Predictive Troubleshooting

By analyzing code and system logs, AI, when prompted correctly, can predict potential bottlenecks or failures. A prompt such as "Analyze the system logs for potential database issues in the next 48 hours" can offer proactive solutions.

Personalized Learning and Skill Development

For budding developers, AI-guided platforms can tailor learning resources. Prompts such as "Provide Python intermediate level problems with solutions" can cater to individual learning curves, fostering more effective skill acquisition.

Ethical Implications and Best Practices

It's essential to ensure that the code generated by AI, guided by prompts, adheres to ethical standards, especially if it's for applications such as data handling or user privacy. Furthermore, developers should review AI-generated code for potential vulnerabilities. For example, if a developer prompts for "Creating a user authentication system," they must ensure that the resulting code follows best practices in security and data protection.

Future Prospects

The integration of AI in software development is still in its nascent stage. As AI models become more sophisticated, their understanding of intricate coding paradigms will improve. Future prompts might be more abstract, such as "Design a system that can handle 1 million concurrent users," with AI providing not just code but a comprehensive design, complete with database schema, server architecture, and potential bottlenecks.

Prompt engineering in the realm of coding and software development marks a paradigm shift. As AI continues to enhance its code comprehension and generation capabilities, the synergy between human developers and AI will become more refined. Carefully crafted prompts will bridge the gap, ensuring that AI tools understand and execute developer intentions accurately. This collaborative approach promises to elevate software design and implementation to unprecedented levels of efficiency and innovation. The coding world is on the cusp of an AI-augmented future, and prompt engineering will be the guiding light in this journey.

Prompts for Entertainment and Gaming

The entertainment and gaming industry has witnessed radical transformations due to advancements in technology. AI's intervention, particularly the integration of sophisticated models through prompt engineering, has opened new avenues for content creation, game design, user engagement, and much more.

Storyline Generation

Through AI, writers and developers can explore myriad story arcs and character developments. A prompt such as "Generate a mystery storyline set in Victorian London" could produce a unique narrative for a game or a TV show, potentially leading to novel franchises.

Character Design and Evolution

Customizing game characters can be enhanced using AI-driven prompts. For instance, "Design a character with steampunk influences, proficient in archery" might result in a detailed character sketch, both in terms of visuals and background.

Dynamic Game Environments

AI can dynamically alter game environments based on player behavior. Prompts such as "Adjust the game's difficulty based on player's skill level" ensures that players are always challenged, enhancing user engagement.

Music and Sound Effects

Original soundtracks and ambient sounds greatly influence a game's atmosphere. By inputting, "Create a haunting melody for a deserted castle level," developers might receive a unique composition tailored for that specific environment.

Dialogue Generation

For games with extensive narratives, AI can be prompted to generate dialogues, keeping in line with character personalities and story arcs. "Generate a conversation between a rogue thief and a city guard" could provide dynamic interactions enhancing the game's realism.

Real-Time Player Feedback

In multiplayer games, AI can facilitate real-time feedback, guiding players. A command such as "Provide hints when a player is stuck in a puzzle for more than 15 minutes" can improve user experience significantly.

Predictive Gaming Trends

Using AI analysis, developers can forecast gaming trends. By prompting "Analyze player data to predict popular gaming genres in 2023," gaming companies can stay ahead of the curve, developing content that aligns with future demands.

Virtual Reality (VR) and Augmented Reality (AR) Experiences

The immersion of VR and AR can be intensified using AI-driven prompts. "Design a VR landscape mimicking 18th century Paris" could craft detailed and historically accurate environments for players to explore.

Ethical Gaming Practices

With the rising concerns about gaming addictions and the potential negative impacts of gaming, AI can be prompted to ensure ethical engagement. A directive such as "Alert players who have been gaming continuously for 4 hours and suggest a break" could promote healthier gaming habits.

Personalized Gaming Experiences

One of the primary advantages of AI is its ability to personalize experiences. By analyzing player data, prompts such as "Adjust storyline based on player's past choices" can offer a gaming experience that's uniquely tailored to each player.

Crowd Control and Online Moderation

For multiplayer games or online platforms, AI can be prompted for moderation tasks. Commands such as "Monitor chat for aggressive language and issue warnings" can ensure a more respectful and inclusive gaming community.

Gamification of Non-gaming Platforms

Entertainment is not limited to traditional games. Streaming platforms, social media, and even educational apps are increasingly integrating gamified elements. Here, AI can be instrumental. A prompt such as "Introduce a scoring system for viewers on a streaming platform based on content watched" can transform passive viewing into an interactive experience.

Merchandising and Ancillary Content

For successful franchises, there's potential beyond the game or show itself. AI can be prompted to design merchandise or even spin-off storylines. "Design a poster based on the popular game character X" or "Draft a prequel storyline for character Y" can lead to diversified revenue streams.

AI's foray into entertainment and gaming is much more than just a futuristic concept—it's the reality of contemporary content creation. Prompt engineering acts as the bridge between human creativity and AI's computational prowess. By guiding AI models through precise, imaginative prompts, developers, writers, and artists can leverage technology to craft content that resonates, engages, and enthrall. As the lines between reality and virtual blur, the entertainment and gaming realm, powered by AI and steered by innovative prompts, stands at the cusp of a renaissance.

Advanced Techniques Personalized and Adaptive Prompts

Prompt engineering is an art as much as it is a science. In the ever-evolving realm of AI, where models become more sophisticated, it's crucial to ensure that prompts can evolve alongside to harness the full potential of these systems. One of the most recent advancements in prompt engineering is the rise of personalized and adaptive prompts that tailor AI outputs to individual users and dynamically adjust based on contexts.

The Need for Personalization and Adaptability

In today's digital age, consumers expect tailored experiences. From personalized shopping recommendations to adaptive learning platforms, the demand for customization is pervasive. This expectation naturally extends to AI-powered applications. A generic AI output is often less impactful than one that is tailored to an individual's preferences, history, or immediate context.

Personalized Prompts—Beyond Generic Queries

By factoring in user-specific data such as past behaviors, preferences, or demographic information, prompts can instruct AI models to generate outputs specifically tailored to an individual. For example, a generic prompt might be "Suggest a book," but a personalized version could be "Suggest a mystery novel for a 28-year-old who enjoyed Agatha Christie's works."

Adaptive Prompts—Responding to Dynamic Contexts

Adaptive prompts adjust based on the real-time context in which they are deployed. If a user is browsing an online store at noon, the prompt could adapt to suggest lunchtime deals. If the same user returns at midnight, the prompt might instead guide the AI to recommend late-night shopping deals.

Balancing Privacy with Personalization

For prompts to be personalized, they require data. It's essential to strike a balance where personalization enhances user experience without infringing on privacy rights. Clear consent mechanisms and transparent data-handling practices are paramount.

Feedback Loops for Adaptive Prompting

To ensure prompts remain adaptive, systems can be designed to include feedback loops where AI learns from user interactions. For instance, if a user consistently ignores or rejects certain types of suggestions, the AI, informed by adaptive prompts, will recalibrate its future recommendations.

Branching Prompts for Multistep Interactions

In applications where user-AI interaction is an ongoing conversation, prompts can be engineered to branch based on prior responses. This creates a dynamic and adaptive flow of interaction, much like a natural conversation.

Real-World Application: E-learning Platforms

Personalized and adaptive prompts are revolutionizing e-learning. An AI tutor, through effective prompt engineering, can offer study material and quizzes tailored to a student's proficiency level and learning pace, dynamically adjusting as the student progresses.

Personalized Health Care Recommendations

In digital health platforms, prompts can be engineered to take into account a user's medical history, age, gender, and other factors, guiding AI models to offer more tailored health advice, workout routines, or diet plans.

Entertainment and Media Consumption

Streaming platforms, through advanced prompting techniques, can offer viewing or listening suggestions that go beyond generic genres, factoring in a user's mood, past viewing history, and even the time of day.

E-commerce and Online Retail

E-shops powered by AI can use personalized prompts to guide models in suggesting products, not just based on a user's browsing history but also considering upcoming occasions, past purchases, and current sale events.

Challenges Ahead

While the prospects of personalized and adaptive prompts are promising, challenges remain. The ethical considerations of data usage, the potential for reinforcing biases, and the technical hurdles in ensuring smooth real-time adaptability are areas that need continuous attention.

Continuous Learning and Evolution

For personalized and adaptive prompts to remain effective, a commitment to continuous learning is essential. As AI models evolve, so should the prompts that guide them, ensuring that the balance between user expectations and AI capabilities is maintained.

Personalized and adaptive prompts represent the future of prompt engineering, merging the vast computational potential of AI with the intricate nuances of individual human experiences. By prioritizing user-centric experiences and understanding the dynamism of real-world contexts, these advanced prompting techniques will play a pivotal role in shaping the next wave of AI applications across industries. As we continue to integrate AI deeper into our daily lives, the emphasis will shift from merely generating AI outputs to generating meaningful, context-aware, and user-specific AI interactions.

Advanced Topics in Prompt Engineering

Table of Contents

In the field of AI, particularly with generative models, prompt engineering has risen as a crucial area, enabling users to effectively communicate with and elicit desired responses from AI. As the bridge between human intent and AI output, prompt design needs to be both effective and adaptive. One of the most potent ways to achieve this adaptability is through the incorporation of user feedback. Here's a detailed exploration of how user feedback can be integrated into prompt design and why it is essential.

The Role of Feedback in Refinement

At the heart of iterative development, be it in software, product design, or AI, lies feedback. In the context of prompt engineering, feedback provides insights into how well the AI is meeting user expectations. When users interact with a prompt and find the output unsatisfactory, their feedback becomes a critical data point for prompt refinement.

Active Listening Mechanisms

AI models equipped with feedback loops can be thought of as active listeners. Just as a person might adjust their responses based on the reactions of their conversation partner, an AI system can modify its outputs based on user feedback, leading to more tailored and relevant results over time.

Feedback-Driven Prompt Libraries

Imagine a repository of prompts that evolves based on user interactions. As users provide feedback, less effective prompts could be modified or phased out, while successful ones could be highlighted or expanded upon. This dynamic library would be a living testament to collective user experience and preferences.

Avoiding Repeated Pitfalls

Certain prompts may consistently lead to undesirable outputs. By incorporating feedback, these prompts can be flagged and adjusted, ensuring that users don't repeatedly encounter the same issues, enhancing user experience significantly.

Encouraging User Participation

By creating systems where users know their feedback directly influences AI behavior, there's an intrinsic motivation to provide constructive input. This active participation can lead to a symbiotic relationship between the user and the AI, fostering an environment of continuous improvement.

Enhancing Trustworthiness

One of the challenges with advanced AI models is the black box nature, where their decision-making processes are opaque. By allowing users to provide feedback and see tangible changes in response, trust in the AI system can be bolstered, as users feel a sense of control and understanding.

Ethical Feedback Integration

While feedback is invaluable, it's also essential to ensure that it doesn't inadvertently introduce biases into the system. For instance, if feedback primarily comes from a specific demographic, the AI might lean toward that group's preferences, sidelining others. Ethical considerations are paramount when incorporating feedback to ensure inclusivity and fairness.

Feedback as a Tool for Education

Sometimes, the gap isn't just in the prompt's design but also in user understanding of AI capabilities. Feedback can be a two-way street. When users consistently face issues, the system, armed with feedback data, can guide users on crafting better prompts, educating them on optimal interaction methods.

Continuous Evolution of AI Systems

Incorporating feedback ensures that the AI system is not static. As users' needs and the external environment change, feedback-driven prompt design ensures that the AI system evolves in tandem, remaining relevant and effective.

Feedback for Diverse Applications

Prompts are used across various domains, from creative writing and coding to business applications. Each domain has unique requirements, and feedback from domain-specific users can help in fine-tuning prompts to better cater to specialized needs.

The dynamic field of prompt engineering stands at the crossroads of AI capabilities and human intent. By integrating user feedback, prompts can become more effective, versatile, and user-centric. Such feedback-driven systems not only enhance the overall user experience but also ensure that as AI technology evolves, it remains anchored to human needs, aspirations, and values. In the ever-evolving landscape of AI, where machines are designed to understand and work with humans, feedback, in essence, becomes the voice of the user, guiding AI systems toward better, more harmonious interactions.

A/B Testing of Prompts

Prompt engineering, a rapidly evolving discipline within the realm of artificial intelligence, revolves around the intricate art and science of communicating intent to an AI model. As AI systems, particularly generative ones, are becoming more ubiquitous, it's imperative that we fine-tune their behavior to best suit user needs. A potent tool at the disposal of prompt engineers to achieve this is A/B testing, a method traditionally associated with web design and user experience optimization. In the context of prompts, it becomes a way to experimentally determine which versions perform best. Here's a detailed dive into the world of A/B testing in prompt engineering

Understanding A/B Testing

At its core, A/B testing involves comparing two versions of a prompt to see which one produces better outcomes. These versions (A and B) are presented to users at random, and the resulting interactions are monitored to determine which version leads to more desired behaviors or outcomes.

Why A/B Testing Is Crucial

In the world of AI, where models can sometimes be black boxes with unpredictable outputs, A/B testing offers a quantitative method to glean insights. It can highlight subtle differences in prompts that lead to significantly varied AI responses.

Setting the Right Metrics

Before running an A/B test, it's crucial to decide on the metrics. Depending on the application, this could range from user satisfaction scores, accuracy of AI-generated content, user engagement levels, or even retention rates for applications that use the AI frequently.

Crafting Variations

The essence of A/B testing lies in the differences between the two prompts. These could be as simple as changing the phrasing or as complex as redesigning the structure of the prompt entirely. Each variation aims to explore a hypothesis about potential improvements.

Sample Size and Duration

A crucial component of A/B testing is ensuring that the sample size is large enough to obtain a statistically significant result. This requires determining the number of users or interactions needed to confidently ascertain a difference between the two prompts. Moreover, the duration of the test needs to be sufficient to capture varied user interactions and account for any potential anomalies.

Analyzing Results

Once data has been collected, the next step is analysis. This involves comparing the performance metrics of the two prompts and determining whether the observed differences are statistically significant. Modern tools and software can aid in this analysis, ensuring robust conclusions.

Iterative Testing

A/B testing isn't a one-off event. It's an iterative process. Based on the results of one test, new hypotheses can be formed, leading to further refinements and subsequent tests. This continuous loop ensures that prompts are always optimized for the current user base and application context.

Potential Pitfalls

While A/B testing offers numerous benefits, there are pitfalls to avoid. Running multiple tests simultaneously without proper segmentation can lead to confounding results. Also, not accounting for external factors, such as a sudden influx of new users, can skew results. It's vital to ensure the testing environment is as controlled as possible.

Beyond Binary Testing

While traditionally A/B testing involves two versions, advanced applications in prompt engineering might necessitate A/B/C testing or even more variations. This allows for a broader exploration of potential improvements but also requires more intricate analysis.

Ethical Considerations

Any form of user testing, including A/B testing, comes with ethical responsibilities. Users should be informed if they are part of a test, and their data should be handled with utmost care, respecting privacy regulations and ensuring anonymity.

A/B testing in prompt engineering is a bridge between the quantitative rigor of experimental science and the qualitative art of crafting effective prompts. By systematically exploring variations, prompt engineers can navigate the vast landscape of possible interactions with an AI, homing in on those that are most effective, user-friendly, and aligned with the application's goals. As AI systems continue to integrate deeper into daily lives and professional environments, such testing methodologies will play a pivotal role in ensuring these systems remain transparent, predictable, and valuable to their users.

Personalized and Adaptive Prompts

The true potential of AI and ML, especially within the context of prompt engineering, lies in the ability to craft responses that are personalized and adaptive to each user. By customizing prompts, systems can better serve users, offering experiences tailored to individual needs and preferences. This not only enhances user experience but also aids in achieving specific business outcomes. Here's an exploration of personalized and adaptive prompts and their significance.

The Rise of Personalization

As the digital world becomes more crowded, businesses are increasingly turning to personalization as a way to stand out. Personalization ensures that each user feels recognized and valued, leading to increased user engagement and loyalty.

What Is a Personalized Prompt?

A personalized prompt is tailored to a specific user based on their unique data, such as browsing history, purchase patterns, or expressed preferences. Instead of generic responses, users receive answers that are directly relevant to them.

The Need for Adaptive Prompts

The digital landscape is dynamic, with user behaviors, preferences, and external conditions continuously changing. Adaptive prompts are designed to adjust in real time to the evolving context, offering optimal responses under varying circumstances.

Data: The Backbone of Personalization

For AI systems to generate personalized prompts, they require access to relevant user data. This can include demographic information, behavioral patterns, interaction histories, and more. The more accurate and comprehensive the data, the better the AI can tailor its responses.

Feedback Loops Are Crucial

Continuously refining and optimizing prompts require real-time feedback loops. As users interact with the system, feedback helps in adjusting and honing the AI's responses, ensuring that the personalization remains effective and relevant.

Dynamic Learning and Prompt Evolution

One of the key features of adaptive prompts is that they evolve. Through ML, these prompts can learn from each interaction, enhancing their capability to respond more effectively over time.

Ethical Considerations

With the power of personalization comes the responsibility of using data ethically. It's crucial to ensure user privacy, obtain necessary permissions for data use, and provide transparency on how data is utilized.

Challenges of Over-Personalization

While personalization offers many benefits, there's a fine line between customization and intrusiveness. Over-personalization can lead users to feel their privacy has been invaded, leading to mistrust. Hence, it's essential to strike a balance.

Multimodal Personalization

In the age of interconnected devices and platforms, personalized prompts aren't limited to text. They span across various modalities, from visual cues and auditory signals to tactile feedback, offering a comprehensive user experience.

Context Is King

The effectiveness of a personalized or adaptive prompt often hinges on the system's ability to understand context. This encompasses not only the immediate query but also broader situational awareness. For instance, a prompt that understands a user is shopping for winter wear during a snowstorm can make more relevant suggestions than one that lacks this context.

Testing and Optimization

Crafting personalized and adaptive prompts isn't a one-time affair. Constant A/B testing, user feedback analysis, and iterative refinement are essential to ensure the system remains effective and resonates with users.

The Broader Implications for Business

Businesses that successfully implement personalized and adaptive prompt engineering can realize significant benefits. These range from increased user engagement and satisfaction to higher conversion rates, customer retention, and overall enhanced brand loyalty.

Future Directions

The future holds promise for even more advanced personalization techniques. We can anticipate systems that understand human emotions, nuances in tone, and other subtleties, crafting prompts that resonate on a deeply personal level.

Personalized and adaptive prompts represent the next frontier in AI-powered interactions. They offer the promise of deeply individualized experiences, resonating with users on a level that generic responses cannot achieve. However, with this power comes the responsibility of ethical and judicious use. By prioritizing user needs, respecting their privacy, and continuously iterating based on feedback, prompt engineers can harness the power of personalization to its fullest potential, crafting experiences that truly elevate user interactions to new heights.

Multilingual Prompt Design

In the age of globalization and digital interconnectedness, the world is becoming a smaller place, but the diversity of languages remains vast. The world speaks thousands of languages, with over 20 major ones having a significant impact on global commerce, culture, and communication. As AI becomes more integrated into our daily routines and business processes, the need for multilingual support becomes increasingly evident. When it comes to prompt engineering, this presents both challenges and opportunities. Here's a detailed look at the intricate world of multilingual prompt design.

The Importance of Multilingualism

The digital world is not English-centric. While English remains a dominant language online, a significant portion of the global online community converses in languages such as Mandarin, Spanish, Arabic,

Hindi, Bengali, and Portuguese. To make AI universally accessible and effective, multilingual prompt design is crucial.

Recognizing Cultural Nuances

Beyond just language, prompts need to account for cultural context. What works in English for a Western audience might not resonate, or worse, might be offensive or misunderstood in another language and culture.

Challenges of Translation

Direct translation often doesn't work. Certain phrases or ways of structuring a sentence in one language might not have direct counterparts in another. The solution requires a combination of linguistic expertise and cultural understanding.

Dialects and Variations

Languages aren't monolithic. For instance, Spanish spoken in Spain differs from that in Mexico or Argentina. Portuguese in Portugal is not the same as in Brazil. Each variation can significantly impact the design of effective prompts.

Designing for Non-Latin Scripts

Languages such as Mandarin, Arabic, and Hindi don't use the Latin alphabet. This introduces unique challenges, especially when considering AI's understanding and generation of these scripts.

Handling Right-to-Left (RTL) Languages

Languages such as Arabic and Hebrew are written from right to left. Designing prompts for RTL languages requires special attention to ensure the AI can correctly interpret and generate content in these languages.

Hybrid Languages and Code-Switching

In many multilingual societies, people switch between languages seamlessly in conversation. Recognizing and appropriately responding to code-switching (e.g. Hinglish, a mix of Hindi and English) is another layer of complexity in multilingual prompt design.

Specialized Vocabulary and Jargon

Certain industries or fields might have jargon that, when translated, doesn't carry the same meaning. For instance, technical or medical terms might not have direct translations in all languages.

Evaluative Metrics across Languages

One of the challenges is how to measure the effectiveness of a prompt in different languages. The criteria for a successful interaction in one language might not be the same in another, especially when considering cultural norms and expectations.

Continuous Improvement and Feedback Loop

Due to the vastness and complexity of languages, it's almost a certainty that initial multilingual prompts will not be perfect. It's essential to have a robust feedback mechanism to continuously refine and improve these prompts based on real-world interactions.

Ethical Considerations

Just as with any application of AI, multilingual prompt design must be approached ethically. There's a responsibility to ensure representation, avoid cultural biases, and ensure that every user, irrespective of their language, receives a respectful and accurate response.

Looking to the Future

With advancements in natural language processing (NLP) and the increasing amount of non-English data available for training, we can expect improvements in multilingual AI capabilities. The next frontier might be not just understanding multiple languages but also understanding the nuances of human emotion and intent across cultures.

Multilingual prompt design is both an art and a science. While the linguistic and technical challenges are significant, they're not insurmountable. With careful attention to detail, a deep understanding of cultures, and continuous feedback, it's possible to create AI systems that communicate effectively across the linguistic tapestry of our world. As businesses and services become more globalized, the importance of this capability will only grow, making it an essential skill in the AI world.

Prompts for Multimodal AI Models

As artificial intelligence continues to evolve, we witness the rise of multimodal models, which are designed to understand and generate content that spans multiple modes or types of data, such as text, images, and sound. This versatility can significantly amplify the AI's capabilities. The challenge, however, lies in effectively prompting such models to leverage their full potential. This article delves into the world of prompts for multimodal AI models, their importance, and strategies for their optimization.

What Are Multimodal AI Models?

Multimodal AI models are engineered to process, interpret, and produce information across various forms or modalities. Instead of being restricted to just text or images, these models can combine different data types to provide richer, more contextual outputs. For example, given an image and a textual description, a multimodal model can generate a relevant story or even an audio narration.

The Complexity of Multimodal Prompting

Prompts for multimodal models aren't as straightforward as those for unimodal models. They often require combining different types of data, ensuring that they align coherently, and determining the right format to elicit the desired response.

Importance of Context

In multimodal AI, context plays a pivotal role. The textual prompt should align with other data types (such as images) to ensure a coherent output. For instance, when describing a sunset image, the textual prompt should provide context that aligns with the visual content, leading to more accurate and relevant responses.

Sequential versus Parallel Prompts

Depending on the task, prompts can be designed to process data sequentially or in parallel. Sequential processing might involve interpreting text and then generating a corresponding image, while parallel processing might involve interpreting text and image simultaneously to produce an output.

Challenges in Crafting Multimodal Prompts

Ensuring coherence and relevance across different data types is challenging. The prompt needs to effectively bridge various modalities while preventing ambiguity, ensuring the model doesn't prioritize one modality over another, unless explicitly instructed to do so.

Exploiting the Strengths of Each Modality

Different modalities offer different strengths. For instance, images can provide a clear visual context, while text can provide nuanced explanations or emotions. Effective prompts will harness the strengths of each modality to produce comprehensive and detailed outputs.

Feedback Loops Are Crucial

Due to the complexity of multimodal AI, iterative feedback becomes even more essential. Continuous testing and refinement of prompts ensure that the model's responses align with the desired outcome, across all included modalities.

Ethical Considerations

Multimodal models, given their capability to interpret diverse data types, have a broader range of applications, potentially intensifying the ethical implications. For example, generating realistic videos with corresponding audio can have implications in deepfake production. Prompt engineers need to be acutely aware of the potential misuses and craft prompts with ethical guidelines in mind.

Training Data and Multimodal Learning

For a multimodal model to effectively respond to prompts, it needs to be trained on diverse and rich datasets that span multiple modalities. The quality of this training data directly impacts the model's ability to understand and generate coherent multimodal outputs.

The Future of Multimodal Prompting

As technology advances, we can anticipate even more integrated models that combine numerous data types, potentially including tactile or sensory information. The future might see prompts that don't just combine text and visuals but also incorporate other human senses, crafting truly immersive AI experiences.

Multimodal AI represents the convergence of different data realms, offering the promise of richer, more detailed AI interactions. The key to unlocking this potential lies in effective prompt engineering, which can navigate the complexities of multiple data types to produce coherent and contextually relevant responses. As we advance in the AI journey, the art of multimodal prompting will undoubtedly play a pivotal role, shaping how we interact with AI in deeply integrated, holistic ways.

Advanced Techniques: Advanced A/B Testing Strategies

Prompt engineering, the art of crafting effective prompts to guide AI responses, is a rapidly evolving field that significantly impacts the efficiency and accuracy of AI systems. As the demand for more nuanced and contextually accurate AI outputs increases, the necessity to refine prompts becomes crucial. As previously explained, A/B testing, or split testing, is an empirical method of comparing two versions against each other to determine which performs better. In the context of prompt engineering, A/B testing is a tool to refine and optimize prompts. This section delves deeper into advanced A/B testing strategies tailored for prompt engineering.

Why Advanced A/B Testing?

At its core, A/B testing involves pitting two different prompts against each other and assessing their performance based on predefined metrics. However, as AI systems grow more complex, and the use cases more diverse, basic A/B testing may not suffice. Advanced strategies aim to tackle these complexities and ensure that the AI system's outputs align closely with desired outcomes.

Multivariate Testing

While traditional A/B testing compares two versions, multivariate testing evaluates multiple variations simultaneously. For prompt engineering, this could mean testing various components of a prompt, such as its structure, wording, or sequence, all at once. This method provides a more holistic view of how different elements interact with each other.

Sequential A/B Testing

Rather than testing two prompts simultaneously, sequential A/B testing evaluates them in succession. This strategy can be particularly useful when assessing prompts that might be influenced by external factors such as time or trending topics.

Dynamic Traffic Allocation

Instead of dividing traffic equally between prompt variations, dynamic traffic allocation uses algorithms to send more traffic to the currently "winning" prompt. It's a more aggressive form of testing that can quickly determine superior prompts but requires vigilant monitoring.

Bandit Testing

Blending the best of exploration (trying out new prompts) and exploitation (using the best-performing prompt), bandit tests allow for continuous learning. This strategy can adapt in real time, providing more immediate optimizations.

Layered Testing

In situations where multiple tests are run concurrently on the same system, layered testing ensures that one test does not unduly influence the outcome of another. This method allows for the simultaneous optimization of multiple prompt components.

Longitudinal Testing

Considering the temporal aspect, longitudinal testing evaluates the performance of prompts over extended periods. It helps in understanding the decay or improvement in a prompt's efficiency over time.

Contextual Bandit Testing

Building on the bandit testing approach, contextual bandit testing incorporates external contextual information, ensuring the AI system's response is not only based on the prompt but also on surrounding data, leading to more contextually accurate outputs.

Personalized A/B Testing

Understanding that one size doesn't fit all, personalized A/B testing involves crafting prompts based on specific user profiles or segments. It evaluates prompt performance within these tailored groups rather than a generalized audience.

Feedback Loop Integration

Advanced A/B testing integrates feedback loops, ensuring that the learnings from testing are immediately fed back into the system for continuous refinement. This dynamic strategy ensures that prompt engineering remains agile and responsive.

Ethical Considerations

While A/B testing aims to improve prompts, it's vital to ensure that tests are conducted ethically. Users should be made aware, data privacy should be maintained, and any potential AI outputs that could be harmful or misleading should be monitored and mitigated.

Challenges in Advanced A/B Testing

Despite its advantages, advanced A/B testing isn't without challenges. Ensuring statistical significance, managing overlapping tests, and interpreting complex results can be daunting. However, with meticulous planning and execution, these challenges can be addressed.

Advanced A/B testing strategies provide a nuanced and comprehensive approach to prompt engineering. As AI systems become integral to more sectors and industries, ensuring their accuracy and relevance becomes paramount. Advanced A/B testing, with its array of sophisticated strategies, offers the tools necessary to hone prompts to perfection. By leveraging these techniques, prompt engineers can ensure that AI systems produce outputs that are not only accurate but also contextually relevant, leading to more meaningful and powerful results.

10

Prompt Engineering with OpenAI ChatGPT

Table of Contents

What Is GPT-4?

Generative Pre-trained Transformer 4, more commonly referred to as ChatGPT-4, is a monumental development in the field of artificial intelligence, specifically in natural language processing (NLP). Developed by OpenAI, ChatGPT is not just another model in a lineage of advancements; it signifies a paradigm shift in how we interact with and understand AI models. With a massive 175 billion parameters—more than two orders of magnitude larger than its predecessor GPT-2—it's one of the most extensive AI models in existence. But ChatGPT is not just about its size; it's about the capabilities that such scale brings forth.

Originating from the transformer architecture, which is fundamentally based on attention mechanisms, ChatGPT-3 leverages this foundational structure to process and generate humanlike text with unprecedented accuracy. The sheer volume of parameters allows the model to store vast amounts of information, akin to a highly intricate digital brain. When we talk about the model's training, ChatGPT-3 has been exposed to a considerable fraction of the Internet. From articles to blogs, books to forums, it has seen diverse forms of human language, enabling it to answer queries, generate narratives, and more with a depth of understanding previously unseen in AI models.

One of ChatGPT-3's standout capabilities is its proficiency in zero-shot, few-shot, and many-shot learning. Instead of needing vast amounts of labeled data for each specific task, ChatGPT-3 can generalize from examples provided in the prompt, adapting its responses based on the context. For instance, if you wanted to translate English to French, you might provide a couple of example translations (few-shot learning) and then present a new sentence, and ChatGPT-3 would attempt the translation based on those examples.

While text generation is its primary function, the applications of ChatGPT-3 stretch far beyond just creating paragraphs of text. Entrepreneurs, developers, and researchers have utilized ChatGPT-3 for tasks such as coding—where it generates snippets based on descriptions—creative writing, tutoring in various subjects, simulating video game characters, and even crafting conversational agents that can engage users meaningfully. OpenAI's decision to provide ChatGPT-3 as an API has further democratized access, letting a broader community of developers integrate ChatGPT-3's capabilities into myriad applications, spawning a renaissance of AI-driven tools and platforms.

However, as with all powerful tools, ChatGPT-3 brings forth ethical and practical challenges. Its capacity to generate content that's nearly indistinguishable from human-written text raises concerns about misinformation, authenticity, and content integrity in the digital age. Moreover, biases inherent in the vast swathes of Internet data it was trained on mean ChatGPT-3 can, at times, produce outputs that reflect societal prejudices, prompting discussions about transparency, control, and responsibility in AI development.

The rise of ChatGPT-3 is emblematic of a broader trend in AI, where models are becoming more generalized, moving away from the era of highly specialized models for each task. As we stand on the cusp of this new age, ChatGPT-3 is a harbinger, a signpost pointing toward a future where the lines between human and machine-generated content blur, where AI becomes an even more integral collaborator in myriad human endeavors, and where the capabilities—and challenges—of AI become central to discussions in tech, policy, and society at large.

How Does ChatGPT-3 Use Prompts?

Basic Working Principle Generative Pre-trained Transformer 3 (ChatGPT-3) is built on the foundation of predicting the next word in a sequence. But this isn't a mere guess; it's a deeply informed prediction based on patterns seen during its training on vast amounts of text.

To make these predictions, ChatGPT-3 leverages layers upon layers of transformers, which are neural network architectures designed to handle sequential data. Each layer processes the input text, capturing various features of the language, from basic grammar to complex contextual relationships. When a prompt is fed to ChatGPT-3, it doesn't simply view it as a static piece of information. Instead, it dives deep into its layers, weighing the importance of each word, and its relationship with other words, to discern patterns and information hierarchies.

This foundation is crucial for any task given to ChatGPT-3, whether it's answering a question, generating a story, or providing a translation. By always being in a mode of predicting the next word, it can generate coherent and contextually appropriate sentences, paragraphs, or even longer pieces of text.

Parsing the Prompt When ChatGPT-3 receives a prompt, its primary goal is to understand the context and the task's intention. Parsing isn't just about recognizing the words in the sequence but about grasping the underlying structure and relationships between those words.

Each word in the prompt acts as a clue. For instance, question words such as "what," "how," and "why" might indicate an informational response is needed. Commands or imperative forms might denote action-based tasks, such as generating a list or a story. Similarly, contextual cues, such as dates, names, or specific terms, provide ChatGPT-3 with anchors to ensure the generated response aligns with the prompt's topic.

Furthermore, ChatGPT-3 doesn't just rely on the immediate context but also taps into broader patterns it has seen during training. For example, if the prompt mentions "Eiffel Tower," even without explicit information, ChatGPT-3 knows it's related to Paris, architecture, and possibly tourism. Such associations help the model generate relevant content even if the prompt is not exhaustive in detail.

However, while ChatGPT-3 is brilliant at recognizing patterns and making connections, it doesn't truly understand the prompt as humans do. It doesn't have beliefs, knowledge, or consciousness. Instead, it predicts responses based on patterns and structures it has seen in its vast training data. The clearer and more explicit a prompt, the better the model can align its response to the user's intention.

Zero-Shot, Few-Shot, and Many-Shot Learning

Zero-shot learning: ChatGPT-3 can attempt tasks without any prior examples in the prompt. For instance, you could ask it to "translate this English text to French: 'Hello, world!'" without prior translations.

Few-shot learning: Here, the prompt contains a few examples to guide ChatGPT-3. If you provide two or three translations and then give a new sentence, ChatGPT-3 uses those examples to deduce the desired output format.

Many-shot learning: This involves providing even more examples in the prompt, refining ChatGPT-3's understanding of the task.

Generating the Response After processing the prompt and any provided examples, ChatGPT-3 begins generating a response. It uses the patterns recognized from the prompt and aligns them with its trained patterns to produce a coherent sequence of words. Each word is chosen based on maximizing the probability of producing a coherent and contextually relevant sentence.

Refining through Iteration If the initial output isn't satisfactory, users can adjust the prompt, make it more explicit, or provide more context. This iterative process is an integral part of effective prompt engineering with ChatGPT-3.

Challenges with Ambiguity Ambiguous or overly broad prompts can lead to generic or off-target responses. For instance, "Tell me about apples" might yield a basic description, while "Describe the nutritional benefits of apples" will produce a more targeted answer. Precision in prompting is essential for desired outputs.

Exploring Different Outputs A unique feature of ChatGPT-3 is that, for a single prompt, it can generate a variety of answers. By slightly rephrasing or introducing randomness into the response generation, users can explore a range of potential outputs for a single input.

Sensitivity to Prompting ChatGPT-3 can produce varied responses based on minute changes in the prompt. This sensitivity can be a double-edged sword. While it allows for granular control over outputs, it can also lead to unintended results if the prompt is not crafted carefully.

Addressing Biases ChatGPT-3's outputs are shaped by its training data. In instances where the data might contain biases, the model can inadvertently reflect them. By understanding and being aware of this, users can craft prompts that guide the model toward more neutral and objective outputs.

Advanced Prompt Engineering As users gain proficiency, they can employ advanced techniques such as chaining prompts, using meta-prompts (prompts that guide the format or style of the answer), or employing external feedback loops to refine the model's output continually.

Conclusion Prompts serve as the interface between the user and ChatGPT-3. Through effective prompting, one can tap into the vast capabilities of the model, from basic information retrieval to complex task execution. Understanding the intricacies of how ChatGPT-3 interacts with prompts is fundamental to harnessing its full potential and opens the door to a myriad of applications across various domains.

Practical Examples of ChatGPT-3 Applications

Content Creation

Description: ChatGPT-3 can be used to generate blog posts, articles, and other forms of written content.

Advantages: It aids in brainstorming, reduces the time needed to draft content, and offers diverse writing styles.

Implementation: By giving the model a topic and desired format, users can get drafts, which can be further refined for publication.

Conversational Agents

Description: Development of chatbots and virtual assistants that can hold natural, humanlike conversations.

Advantages: Enables 24/7 customer support, offers consistent and accurate information, and can be integrated across platforms.

Implementation: ChatGPT-3 can be embedded in websites, apps, or customer service platforms to engage with users in real time.

Code Writing

Description: ChatGPT-3 can generate or complete snippets of code based on descriptive prompts.

Advantages: Assists developers by reducing manual coding, offers solutions to coding challenges, and speeds up software development.

Implementation: Developers can integrate ChatGPT-3 into their IDEs (integrated development environment) or leverage it in coding platforms.

Language Translation

Description: Translating text from one language to another.

Advantages: Provides quick translations, especially for commonly spoken languages, and aids in breaking down language barriers.

Implementation: Users simply input text in one language and specify the desired output language.

Tutoring and Education

Description: ChatGPT-3 can assist students in various subjects by answering questions or explaining concepts.

Advantages: Personalized learning experiences, 24/7 availability, and a wide range of subjects.

Implementation: Can be integrated into e-learning platforms, apps, or educational websites.

Creative Writing

Description: Generates poems, stories, and other creative content.

Advantages: Provides inspiration, develops diverse narratives, and assists in brainstorming sessions.

Implementation: Users can provide themes, opening lines, or characters to guide the model's creative output.

Business Analytics

Description: ChatGPT-3 can interpret complex datasets and provide summaries or insights.

Advantages: Reduces the manual effort of data analysis, provides quick insights, and aids in decision-making.

Implementation: Companies can feed data summaries to ChatGPT-3, asking for trend analysis or predictions.

Gaming

Description: ChatGPT-3 can be used to design in-game dialogues, storylines, and character behaviors.

Advantages: Enhances game narratives, provides dynamic user experiences, and reduces game development time.

Implementation: Game developers can integrate ChatGPT-3 into game engines to produce real-time dialogues or scenarios.

Health Care Assistance

Description: Providing general health information or answering medical queries.

Advantages: Aids in patient education, provides immediate responses, and supports health care professionals.

Implementation: Can be incorporated into health apps, websites, or telemedicine platforms, with the caution that it doesn't replace professional medical advice.

Legal and Compliance

Description: Assisting in legal research, document review, and general legal queries.

Advantages: Speeds up legal research, reduces manual review hours, and provides quick legal information.

Implementation: Legal firms can integrate ChatGPT-3 into their databases or use it to assist clients on websites.

Product Descriptions

Description: Generates product descriptions for e-commerce platforms.

Advantages: Consistency in descriptions, scalability for large product lists, and potential for personalized descriptions.

Implementation: E-commerce platforms can use ChatGPT-3 to generate or refine product details based on key features provided.

Mental Health Support

Description: ChatGPT-3 can provide immediate, empathetic responses, acting as a first-line mental health chatbot.

Advantages: 24/7 availability, immediate support, and nonjudgmental interactions.

Implementation: Can be integrated into mental health apps or websites, with clear disclaimers about professional interventions.

Limitations and Ethical Considerations

The awe-inspiring capabilities of ChatGPT-3, while revolutionary, come with their own set of limitations and a host of ethical considerations. Understanding these is paramount for developers, businesses, and end users who wish to deploy and interact with this technology.

Starting with the limitations, one of the primary constraints of ChatGPT-3, like other models of its kind, is that it lacks a genuine understanding of the content it generates. While it can produce text that appears coherent and contextually relevant, the model does not possess true comprehension. This absence of semantic understanding means that while ChatGPT-3 can generate correct and insightful answers, it can also produce content that is misleading or factually incorrect. This limitation underscores the importance of human oversight, especially in critical applications where accuracy is non-negotiable.

Furthermore, ChatGPT-3 operates based on patterns in the data it was trained on. This means that if it encounters novel or out-of-distribution prompts, there's no guarantee it will respond appropriately. This limitation can lead to unpredictable outputs, potentially causing issues in applications that require consistency.

Another limitation lies in its verbosity. ChatGPT-3 often generates longer responses than necessary, which might not always be suitable for applications requiring concise answers. Additionally, while the model can simulate creativity, it essentially remixes existing information, making it less effective for truly groundbreaking or novel creative endeavors.

On the ethical front, the very strength of ChatGPT-3, its immense training data, poses challenges. Since ChatGPT-3 is trained on vast portions of the Internet, it inadvertently learns and can propagate the biases present in these data. This bias can manifest in outputs that might be considered sexist, racist, or generally prejudiced. Such biases not only pose ethical concerns but can also lead to PR disasters for businesses or misleading information for end users.

Relatedly, ChatGPT-3's capability to generate humanlike text raises concerns about misinformation. In an era where fake news is a significant concern, tools that can produce believable but fabricated narratives can be weaponized to mislead and manipulate audiences. This capability underscores the need for stringent controls on how such models are accessed and used.

Another ethical quandary is the potential loss of jobs. As ChatGPT-3 can generate content, code, answer queries, and more, there's an inherent risk of it displacing jobs across various sectors. While it can act as an augmentation tool, enhancing human capabilities, the thin line between augmentation and replacement is a matter of ongoing debate.

Moreover, ChatGPT-3's conversational prowess leads to concerns about privacy. As users interact with ChatGPT-3-powered applications, they might inadvertently share personal or sensitive information. While OpenAI ensures that interactions with the model are not stored, applications built on top of ChatGPT-3 might not always follow the same protocols, posing potential privacy risks.

Lastly, the democratization of such powerful technology can be both a boon and a bane. While it provides developers across the globe the ability to create transformative applications, it also means malicious actors can harness the model for nefarious purposes. The balance between accessibility and control is a delicate one.

In conclusion, while ChatGPT-3 is undeniably a marvel of modern AI research, its deployment is not devoid of challenges. Recognizing its limitations helps in ensuring it's used appropriately, and being cognizant of the ethical ramifications is crucial for responsible development and usage. As with any powerful tool, the impact of ChatGPT-3 on society is shaped by the choices of those who wield it.

Future Developments and Opportunities

Greater Model Fine-Tuning As OpenAI continues its research, we can expect more sophisticated versions of GPT models that allow for intricate fine-tuning. This will enable developers to specialize the AI for specific tasks, enhancing performance in niche areas.

Real-Time Adaptation Future iterations may incorporate real-time learning mechanisms, adjusting and refining responses based on ongoing user feedback. This will make the AI more contextually aware and adaptive to user-specific needs over time.

Multimodal Functionality Future versions of GPT models might integrate multimodal functionalities, processing both text and other data types such as images or sounds. This will broaden the scope of applications, from rich multimedia content generation to more holistic data analysis.

Integration with Augmented Reality (AR) and Virtual Reality (VR) ChatGPT-3 could be used to generate real-time dialogues or narratives within AR and VR environments. Imagine interactive storytelling where the narrative shifts based on user choices, all orchestrated by the AI.

Automated Software Development With ChatGPT-3's already demonstrated ability to write code based on prompts, future enhancements might lead to fully automated software development for certain tasks, significantly speeding up the development process.

Personalized Learning ChatGPT-3 can be further refined for educational purposes, creating personalized learning experiences. It can adjust its teaching methods based on a student's learning style, pace, and current knowledge level.

Advanced Sentiment Analysis By refining prompt engineering further, ChatGPT-3 could be employed to derive deeper insights from vast textual datasets. This isn't just about determining if a comment is positive or negative, but understanding nuance, sarcasm, and cultural context.

Enhanced Creativity Tools For writers, artists, and other creatives, ChatGPT-3 can act as a collaborative partner, offering ideas, critiques, or even generating drafts. With more refined prompts, the creative process can be augmented in unprecedented ways.

Broader Language Support While ChatGPT-3 already supports multiple languages, there's always room for expansion. Future developments might include better support for regional dialects or even dead languages, enabling diverse and inclusive applications.

Reduced Model Size One of the criticisms of ChatGPT-3 is its sheer size. Future research might focus on creating more compact models without compromising capability. This can make ChatGPT-3 more accessible and cost-effective for smaller developers.

Ethical and Bias Checks Given the ethical concerns associated with AI, we can anticipate the development of integrated tools within ChatGPT-3 that monitor and alert for potential bias in its outputs. This ensures more ethically aligned content generation.

Advanced Gaming Applications The gaming industry stands to benefit immensely. Imagine games where NPC (non-player character) dialogues are not pre-scripted but generated in real time by ChatGPT-3, based on player actions.

Robust Content Filtering As ChatGPT-3 is utilized more for content generation on public platforms, there will be a need for advanced filtering mechanisms to ensure the content adheres to platform guidelines and is free from inappropriate or harmful content.

Integration with IoT Devices The Internet of Things (IoT) ecosystem, filled with smart devices, can leverage ChatGPT-3 for natural language processing, allowing for more intuitive user-device interactions.

Collaboration with Other AI Systems ChatGPT-3 could be paired with other AI systems, such as those designed for image recognition or data analysis, to offer comprehensive solutions. For instance, analyzing visual data and then using ChatGPT-3 to generate a detailed report.

As ChatGPT-3 and its successors continue to evolve, the horizon of opportunities expands. The integration of these advancements will shape a future where AI doesn't just assist but collaboratively works alongside humans in diverse fields. The key will be to harness these capabilities responsibly, ensuring technological advancements align with ethical considerations.

11

Exploring Prompts with ChatGPT

Table of Contents

Introduction to ChatGPT

ChatGPT is a conversational agent developed by OpenAI based on the powerful Generative Pre-trained Transformer (GPT) architecture. It's one of the manifestations of OpenAI's pursuit to design and refine AI models for natural language understanding and generation.

The genesis of ChatGPT lies in the transformative potential that GPT models hold. While the initial versions of GPT were primarily showcased through tasks such as text generation, completion, and translation, the demand for a more interactive and engaging interface was evident. Thus, ChatGPT emerged, tailored to simulate humanlike conversation in real time, responding intelligently to a myriad of user queries and inputs.

Unlike many chatbots that operate based on predefined scripts or limited decision trees, ChatGPT employs a deep learning approach. It was trained on vast amounts of text data, enabling it to generate coherent, contextually relevant, and often creative replies. As with any model, the quality of the response from ChatGPT largely depends on the prompt it receives. Hence, the art of prompt engineering becomes essential, not just to obtain a valid answer, but to guide the model to produce a desired type of response, be it verbose, concise, humorous, or formal.

Underpinning ChatGPT's design is the idea that human-AI interaction should feel intuitive and seamless. As users interact with ChatGPT, they're not just sending queries to a database. Instead, they are engaging in a dynamic dialogue with an AI system that can adapt and respond in real time. This flexibility has led to ChatGPT's adoption in diverse applications, from simple Q&A sessions, tutoring in various subjects, brainstorming ideas, to even simulating characters for video games.

However, it's crucial to understand that while ChatGPT strives to generate accurate and contextually appropriate responses, it's not infallible. The model's responses are predictions based on its training data and the prompts it receives. Thus, it can occasionally produce answers that are incorrect, ambiguous, or even inappropriate. This underscores the need for users and developers to approach ChatGPT with a sense of discernment, recognizing its vast capabilities while also being aware of its limitations.

Another fascinating aspect of ChatGPT is its ability to simulate different personalities or tones based on prompt engineering. With the right prompts, one can make ChatGPT sound like a Shakespearean character, a technical expert, or even a playful jester. This versatility amplifies its utility across various domains, from entertainment to education and beyond.

In summation, ChatGPT represents a significant step forward in the world of conversational AI. It amalgamates the power of GPT models with the interactive essence of chat interfaces, creating a platform where users can engage, inquire, and even be entertained. As with any technology, it's a tool—its efficacy and safety lie in how it's used. With continued research, ethical considerations, and user feedback, ChatGPT and similar models can pave the way for more enriched human-AI interactions in the future.

How Prompts Play a Role in ChatGPT

Foundation of Interaction

- At its core, ChatGPT relies on prompts to initiate any interaction. A prompt is essentially an input that cues the model to generate a specific type of output.
- Unlike simple command-based systems, ChatGPT needs textual prompts to understand context and produce relevant responses.

Contextual Understanding

- Prompts provide the necessary context. For instance, asking, "What's the capital of France?" provides a clear context for a factual answer.
- With more complex prompts, such as "Imagine if Shakespeare was a modern playwright. How would he describe a city?" ChatGPT uses its training to generate creative, contextually relevant replies.

Guiding Response Behavior

- The design of the prompt can steer ChatGPT's responses. A succinct prompt might get a straightforward answer, while a detailed one can lead to a more expansive reply.
- For example, "Tell me about black holes" might yield a general overview, while "Explain the event horizon of a black hole" will produce a more specific explanation.

Tonality and Style

- By tailoring prompts, users can influence the tone of ChatGPT's output. Asking it to "describe a rainforest like a poet" versus "Provide a scientific description of a rainforest" will generate drastically different styles of responses.
- This flexibility is due to the model's vast training data, which encompasses various writing styles and tones.

Continuous Conversation Flow

- For extended interactions, prompts can be sequenced to maintain a conversation flow. Users can pick up from where the last prompt left off, ensuring continuity in dialogue.
- For instance, after getting a response about the solar system, a follow-up prompt such as "Tell me more about Mars specifically" can guide the conversation in a specific direction.

Handling Ambiguities

- Ambiguous or unclear prompts can lead to generic or unexpected responses. Being explicit can help in obtaining more accurate and contextually apt answers.
- For example, "Tell me about Apple" might lead to information about the fruit or the tech company, depending on the model's prediction. Specifying "Apple Inc.'s history" narrows down the context.

Managing Inaccuracies and Biases

- The model's knowledge and potential biases stem from its training data. Specific prompts can be used to probe and understand these biases or to ensure that answers are as neutral as possible.
- For example, using prompts that ask for multiple perspectives on a controversial topic can help in presenting a balanced view.

Simulating Roles and Characters

- By leveraging prompts, ChatGPT can be made to simulate specific roles or fictional characters. For instance, it can be prompted to answer as if it were a historical figure or a famous literary character.
- This has applications in gaming, entertainment, and education, where dynamic character interactions can be beneficial.

Educational Applications

- Prompts play a crucial role when ChatGPT is employed in educational settings. Tutors or educators can craft prompts that guide the model to provide explanations, solve problems, or even generate quiz questions for students.
- For instance, "Explain photosynthesis in simple terms" can be used to get a layman's explanation for young students.

Limiting Response Length and Complexity

- Prompt engineering can dictate the length and complexity of ChatGPT's responses. By specifying constraints, users can obtain short summaries, detailed explanations, or even one-word answers. Example: "In one sentence, summarize the theory of relativity."

Feedback Loop

- Continual interactions with ChatGPT can be seen as a feedback loop. Analyzing responses and adjusting prompts in real time helps in refining the quality and relevance of the conversation.
- It's a dynamic process where users learn to formulate better prompts based on the AI's outputs and vice versa.

Safety and Moderation

- While ChatGPT is designed to avoid harmful or inappropriate content, the specifics of a prompt can sometimes elicit unexpected outputs. Recognizing this, users and developers can craft prompts that minimize the chances of undesirable responses.

In essence, prompts are not just passive inputs for ChatGPT; they are instrumental in shaping the interaction, guiding the AI's behavior, and ensuring that the outputs are aligned with the user's intentions and needs. Proper understanding and crafting of prompts can unlock the full potential of ChatGPT, facilitating rich and meaningful dialogues.

Use Cases and Examples

Content Creation Example: Using ChatGPT to generate blog intros, taglines, or creative stories by prompting, "Write a suspenseful introduction for a mystery novel."

Educational Assistance Example: Asking ChatGPT to explain complex mathematical problems or historical events, e.g. "Can you help me understand the Pythagorean theorem?"

Coding Help Example: Requesting programming solutions or debugging advice, e.g. "Provide a Python script to find the factorial of a number."

Language Translation and Learning Example: "Translate 'Hello, how are you?' into French." Or "Help me practice Spanish. Let's have a basic conversation."

Business and Market Analyses Example: "Summarize the key trends in the e-commerce market in 2020."

Medical and Health Information Example: "Describe the symptoms of the common cold." (Note: Always consult professionals for medical advice.)

Simulation of Characters for Gaming and Entertainment Example: "You are a medieval knight. Describe your day."

Mental Health and Well-Being Support Example: "Provide relaxation techniques for stress." (Note: ChatGPT isn't a substitute for professional mental health support.)

Creative Exercises Example: "Write a poem about rain." Or "Describe a world where trees can talk."

Technical and Scientific Explanations Example: "Explain Einstein's theory of relativity in simple terms."

Roleplaying and Scenarios Example: "Imagine you're an astronaut on Mars. Describe the landscape."

Law and Legal Information Example" "Explain the First Amendment rights." (Note: Always consult a lawyer for legal advice.)

Travel and Geography Insights Example: "Describe the main attractions in Paris."

Cooking and Culinary Guidance Example: "Provide a simple recipe for chocolate chip cookies."

Trivia and General Knowledge Example: "Who won the Oscar for Best Director in 2019?"

Philosophical and Ethical Discussions Example: "Discuss the philosophical implications of AI."

DIY and Home Solutions Example: "How can I fix a leaky faucet?"

Fashion and Beauty Tips Example: "Suggest a summer wardrobe for tropical climates."

Book and Movie Recommendations Example: "Recommend classic novels for a reading list."

Personalized Fitness Regimens Example: "Design a 30-minute workout routine for beginners." (Note: Always consult professionals for fitness advice.)

Brainstorming and Idea Generation Example: "I'm writing a sci-fi novel. Can you brainstorm potential plot points?"

Cultural and Historical Insights Example: "Describe the Renaissance period in Europe."

Financial and Economic Explanations Example: "Explain the concept of inflation."

Mock Interviews and Training Example: "Conduct a mock job interview for a software engineer position."

Relationship and Social Advice Example: "Provide tips for effective communication in relationships." (Note: Always consult professionals for relationship advice.)

Music and Artistic Creation Example: "Describe the evolution of jazz music."

Current Events and News Summaries Example: "Summarize the key global events of 2020."

Work and Productivity Hacks Example: "Give tips to enhance productivity while working from home."

Environmental and Sustainability Information Example: "Discuss the impact of plastic pollution on marine life."

Fun and Games Example: "Tell me a joke." Or "Let's play a word association game."

By leveraging the versatile capabilities of ChatGPT and crafting thoughtful prompts, users can navigate a plethora of scenarios, making the tool an invaluable asset for learning, creativity, and information.

Challenges and Areas of Improvement

Model Bias and Ethical Considerations

One of the primary concerns with models such as ChatGPT is the potential for it to exhibit biases, as it is trained on vast amounts of Internet text. The model may inadvertently reinforce harmful stereotypes or produce content that isn't objective. It becomes crucial to ensure that biases are minimized and that there's a mechanism for users to report and correct misleading outputs.

Specificity and Vagueness in Responses

Depending on how a prompt is phrased, ChatGPT might produce responses that are either too specific or too vague. This could be improved by refining the prompt engineering mechanism to allow for dynamic granularity based on the user's requirements.

Memory Limitations

ChatGPT, especially earlier versions, has a limited token (word or character sequence) memory. This means it might not remember the beginning of a long conversation, which can lead to inconsistencies in replies. Incorporating better short-term memory functions could improve continuity and context-awareness in longer interactions.

Over-Reliance on Certain Patterns

Sometimes, when uncertain, the model may generate overly generalized or "safe" responses. This is because it tends to rely on patterns that were frequent in its training data. Advanced techniques could be developed to push the model to generate more creative or unique outputs when desired.

Difficulty with Ambiguous Prompts

Ambiguity can confuse ChatGPT. While humans might ask questions with implied context, the model requires clear and unambiguous prompts to produce the most accurate responses. Feedback loops and model retraining could improve its handling of ambiguous queries.

Potential for Misinformation

Given the vastness of its training data, there's a possibility that ChatGPT might provide information that's outdated, incorrect, or misleading. Regular updates and incorporating a real-time fact-checking mechanism might help alleviate this issue.

Emotion and Sentiment Understanding

ChatGPT doesn't possess emotions and might struggle to fully grasp and respond to emotionally charged prompts in the way a human would. Enhancing sentiment analysis capabilities can lead to more empathetic and context-aware responses.

Inappropriate or Harmful Outputs

There are concerns about the model generating harmful or inappropriate content, even if unintentionally. A robust moderation layer, perhaps aided by human reviewers, could help ensure that generated content adheres to community and ethical standards.

Dependency and Over-reliance by Users

Users might become too dependent on ChatGPT for tasks, potentially compromising critical thinking and creativity. Educating users about the model's strengths and limitations can foster a more balanced and informed use.

Interactivity and Real-Time Adaptability

The model's responses are largely stateless, meaning it doesn't adapt in real time based on user reactions or feedback. Future iterations could incorporate more interactive and adaptive mechanisms to fine-tune responses based on real-time user feedback.

Resource Intensiveness

Large models such as ChatGPT can be resource-intensive, requiring significant computational power, especially for real-time applications. Optimizing the model for different platforms, including low-power devices, would increase its accessibility and usability.

Handling Multimodal Data

While ChatGPT is primarily text-based, there's a growing demand for models that can handle multiple forms of data (e.g. images, videos). Incorporating multimodal capabilities can make interactions more dynamic and comprehensive.

Cost and Accessibility

Using ChatGPT, especially for high volumes of queries, can be costly for developers. Efforts to make such models more affordable and accessible can democratize AI benefits across different sectors and user groups.

Integration with Other Systems

Integrating ChatGPT seamlessly with other systems, applications, or databases can sometimes be challenging. Improving the model's compatibility and offering a wider range of integration tools can enhance its applicability.

As AI and language models continue to evolve, addressing these challenges becomes imperative. By recognizing these areas of improvement, developers and the broader AI community can work collaboratively to refine ChatGPT and similar models, ensuring they are both powerful and responsible tools for the future.

ChatGPT: The Next Steps

The advent of ChatGPT, powered by OpenAI's GPT models, has revolutionized the landscape of natural language processing. Its ability to comprehend and generate humanlike text has led to a myriad of applications, from customer support bots to creative writing assistants. Yet, as with all pioneering technologies, the journey of ChatGPT is an ongoing one. Let's delve into the anticipated directions, enhancements, and evolution paths that might shape the future of ChatGPT.

Enhanced Contextual Memory

ChatGPT, despite its prowess, has limitations in its ability to retain and reference previous interactions in extended conversations. Future iterations are likely to place significant emphasis on endowing the model with a more robust contextual memory. This would mean that, in extended interactions, the model could maintain consistency and build on previous inputs, fostering a more natural conversation flow.

Reducing Bias and Ethical Reinforcement

Bias mitigation remains a paramount concern. While OpenAI has invested considerable effort in curbing unintended biases in ChatGPT, there's still room for progress. We can expect future versions to be trained with even more stringent guidelines, possibly coupled with real-time bias-detection algorithms, to further ensure objectivity and fairness in outputs.

Fine-Tuned Personalization

Imagine a ChatGPT that knows your writing style or understands your business jargon. Personalization will likely be a significant focus, allowing users to "train" or "tune" ChatGPT on specific data or preferences. This would allow the model to align more closely with individual or enterprise needs, making interactions more relevant and precise.

Integration with Multimodal Models

The realm of AI isn't limited to text. There's an increasing emphasis on multimodal models, which can understand and generate multiple types of data, such as images or audio. The integration of ChatGPT with such models could lead to more dynamic interactions, such as describing images, transcribing audio, or even generating multimedia content based on textual prompts.

Improved Real-Time Adaptability

The vision is to have a model that learns and adapts during the conversation, adjusting its responses based on user feedback or reactions. This real-time adaptability would make interactions with ChatGPT more organic, as it dynamically refines its outputs to align with user expectations.

Broader Domain Specializations

While ChatGPT is a generalist, we might see specialized versions tailored for specific domains, such as medicine, law, or engineering. These specialized models would possess a deeper understanding of their respective domains, offering expert-level insights and interactions.

User-Friendly Interfaces and Tool Kits

To make prompt engineering and model interactions more accessible, we can anticipate the development of more user-friendly interfaces. These would allow even those without deep technical expertise to craft effective prompts, fine-tune model behaviors, and extract maximum value from ChatGPT.

Enhanced Collaboration with Human Intelligence

There's an increasing realization that the best outcomes often arise from a synergy between human and artificial intelligence. Future versions of ChatGPT might be designed to work in tandem with human experts, collaboratively producing content, solving problems, or even brainstorming ideas.

Addressing Resource Efficiency

Highly sophisticated models often come with computational costs. Efforts will likely be directed toward making ChatGPT more resource-efficient, ensuring it remains responsive in real-time applications and can be deployed even on devices with limited computational power.

Community-Driven Evolution

OpenAI has always emphasized community feedback. By continually integrating inputs from developers, users, and researchers, ChatGPT's evolution will likely be a collaborative effort, shaped by the needs, ethical considerations, and aspirations of its vast user base.

While ChatGPT has already made a mark in the annals of AI history, its journey is far from complete. As we stand on the cusp of further advancements in AI, ChatGPT's potential to transform our interaction with machines is vast. By addressing its current challenges and pushing the boundaries of what's possible, the future of ChatGPT holds promise and intrigue in equal measure.

12

Getting Creative with DALL-E

Table of Contents

The Concept of DALL-E

OpenAI, renowned for its advancements in the AI domain, has been at the forefront of introducing technologies that constantly push the boundaries of what machine learning (ML) models can achieve. DALL-E, a derivative of the GPT-3 model, is a testament to OpenAI's innovative approach. Unlike GPT-3, which is tailored to process and generate human language, DALL-E's forte lies in the field of generating images. The name "DALL-E" itself is a playful homage to the surrealist artist Salvador Dalí and the Pixar character WALL-E, hinting at the blend of art and tech that the model encapsulates.

DALL-E is designed to generate images from textual descriptions. To be precise, it's a 12-billion parameter version of the GPT-3 transformer trained to generate images from textual prompts. What this implies is that you can provide a textual description, as abstract or concrete as you want it to be, and DALL-E will attempt to produce an image that matches that description. For instance, prompt it with "a two-headed flamingo" or "a futuristic city skyline," and DALL-E will conjure up unique images that embody those descriptions.

The underlying concept here is more profound than just image generation. It's about bridging the gap between linguistic semantics and visual representation. Traditional image generation models required vast datasets of labeled images to function with accuracy. DALL-E revolutionizes this by leveraging a neural network that understands context, making it possible to generate images of objects or scenarios that don't exist in the real world.

But what makes DALL-E truly intriguing is its ability to handle abstract concepts and juxtapositions that haven't been seen before. It can merge different objects, play with styles and themes, or even create entirely new, fictional entities. The underlying training data, although vast, doesn't contain explicit instances of many of the prompts users might provide. This means DALL-E isn't just regurgitating learned images but creatively interpreting the textual prompts.

Furthermore, DALL-E's prowess extends beyond static images. It exhibits a nuanced understanding of art styles, techniques, and even emotions. Want an image of a sad-looking cube in the style of Picasso? DALL-E strives to capture not only the physicality of the prompt (the cube) but also the emotion (sad-looking) and the artistic essence (Picasso's style).

Of course, the introduction of DALL-E has brought forth discussions on the implications of such technology. While the immediate excitement revolves around the infinite creative possibilities, such as aiding graphic designers, artists, and even filmmakers, there are broader questions to ponder. What does it mean for intellectual property rights if a machine can conjure unique artwork on a whim? How do we ensure the responsible use of such technology?

In conclusion, DALL-E represents a significant leap in the convergence of language understanding and visual creativity within the realm of AI. As with all groundbreaking innovations, it heralds a future filled with possibilities while also presenting new challenges. As the technology continues to evolve, so will our understanding of its potential applications and implications.

How DALL-E Utilizes Prompts

DALL-E, as a revolutionary image-generation model, hinges on the power of textual prompts to bring ideas to visual life. At its core, DALL-E's ability to convert a string of text into a corresponding image is nothing short of transformative. Here's a step-by-step breakdown of how DALL-E utilizes prompts to generate images.

Receiving the Textual Input

The process begins with DALL-E receiving a textual prompt. This can range from simple, concrete concepts such as "apple" to more abstract, imaginative ones such as "a skyscraper shaped like an avocado."

Tokenization of the Prompt

Like many neural network models, DALL-E doesn't understand text in the way humans do. Instead, the provided prompt is tokenized, breaking it down into chunks that the model can process. Each token can represent a word or part of a word, translating human language into a format the machine understands.

Embedding the Tokens

Once tokenized, each token is transformed into a high-dimensional vector using embeddings. These vectors carry semantic information about the tokens, and they play a crucial role in determining the generated image's attributes and features.

Neural Network Processing

DALL-E's architecture, which is based on the transformer neural network, processes these embedded tokens. Through a series of layers and attention mechanisms, the model factors in relationships and dependencies between different tokens to understand the overall context of the prompt.

Decoding and Image Generation

After processing the tokens, DALL-E decodes the resultant information to begin the image generation phase. It starts from a coarse image and refines it iteratively. Pixel by pixel and layer by layer, the image takes shape, guided by the semantic information extracted from the original prompt.

Postprocessing and Refinement

Once an initial image has been produced, there's often a phase of post-processing and refinement. DALL-E might adjust colors, shapes, and other features to ensure the generated image is as coherent and contextually relevant as possible.

Quality Assessment

DALL-E might internally evaluate the quality or relevance of the generated image to the prompt. Though it doesn't possess humanlike judgment, certain built-in metrics can help the model ascertain if the image is likely to be a good representation of the input text.

Output Presentation

The final image, which started as a mere textual description, is then presented as the output. For certain applications or platforms, DALL-E can produce multiple variants of the image, giving users a range of visual interpretations of their prompt.

Iterative Feedback Loop (for Continuous Training)

In advanced implementations, there might be a feedback mechanism. If users can provide feedback on the generated images, DALL-E can learn and improve over time, refining its understanding of prompts and improving the accuracy and creativity of its outputs.

Adapting to Specialized Prompts

While DALL-E is versatile, it's also been observed to be adaptable to specialized or niche prompts. If consistently fed with a certain type of prompt (e.g. architectural designs, fantasy creatures), the model can, over time, fine-tune its outputs to excel in that specific domain.

DALL-E's ability to utilize prompts and transform them into visual artworks is a testament to the leaps AI has taken in recent years. By understanding the intricacies of this process, one gains a deeper appreciation of the blend of art and science that goes into every image this remarkable model generates.

Examples of Generated Artwork

DALL-E, by OpenAI, demonstrated a compelling capability to generate diverse images from text prompts. Some of the real examples from OpenAI's showcase and the subsequent online community experiments are as follows:

"An armchair in the shape of an avocado": This iconic representation demonstrates DALL-E's ability to merge two unrelated concepts. The resulting image is a creatively designed armchair with avocado features, including the pit as a cushion.

"A two-headed flamingo": The model generated a creature with the body of a flamingo and two heads, bringing to life an imaginative and nonexistent entity.

"A storefront that has the word 'OpenAI' written on it": The image depicted a typical storefront but with OpenAI branding, showcasing the model's capability to integrate specific text into scenes.

"A professional high-quality illustration of a futuristic city at night": The result was a detailed urban environment with skyscrapers illuminated with neon lights, emphasizing DALL-E's talent in detailed scene generation.

"A cube of gelatin": Demonstrating DALL-E's capability to visualize simple objects, it produced a perfect translucent cube that resembled gelatin.

"A cross-section of a submarine in a turtle": This whimsical prompt led to an image where a turtle's interior was envisioned as a submarine's inner workings.

"A hexagonal green and yellow clock": DALL-E produced a clock design fitting the given shape and color criteria.

"A teddy bear made of spaghetti": Combining the form of a teddy bear and the texture of spaghetti, DALL-E generated an imaginative creature that seems to be crafted from pasta.

"The Mona Lisa as a Picasso painting": Here, DALL-E reinterpreted the classic Mona Lisa portrait in the abstract, fragmented style reminiscent of Picasso.

"A cyberpunk doctor": The result was a futuristic medic equipped with cybernetic enhancements and a dystopian aesthetic.

"A robotic giraffe": DALL-E combined machinery with the form of a giraffe, resulting in a mechanical version of the creature.

"A futuristic sneaker with wings": The image portrayed a stylish sneaker equipped with winged attachments, hinting at flight capabilities.

"A squirrel holding a cup of coffee": A cute representation, the model produced an image of a squirrel, cup in hand, seemingly enjoying a coffee break.

"A 3D-printed skyscraper on the moon": Showcasing its ability to create fantastical landscapes, DALL-E visualized a lunar setting with an architecturally unique skyscraper.

"A floating vegetable island": The output was a serene scene with an island made up of giant vegetables, floating amid a backdrop of clouds.

These real examples underscore DALL-E's profound potential in digital art, design, and conceptual visualization. By understanding context, combining unrelated ideas, and adhering to specific constraints, it opens the door to vast artistic and commercial possibilities.

Limitations of Dall-E

Computational Costs

- **Resource intensive:** Training models such as DALL-E require vast computational resources, which might be out of reach for most individuals or small entities.
- **Environmental concerns:** The carbon footprint of training such extensive models has raised environmental concerns.

Generalization Limitations

- **Niche requests:** While DALL-E excels at many prompts, there can be very niche or abstract requests where it might not generate appropriate or expected outputs.
- **Overfitting to popular culture:** There's a possibility that the model can be biased toward more popular or dominant cultural imagery because of the data it was trained on.

Ethical and Societal Implications

- **Deepfakes:** The potential misuse for creating realistic and misleading images (deepfakes) can have ramifications in misinformation campaigns.
- **Intellectual property:** Automated generation of artwork could pose challenges to copyright norms and the valuation of human-created art.
- **Economic impacts:** With the ability to produce vast amounts of visual content quickly, there are concerns about potential job losses in design and other creative fields.

Reliability and Predictability

- **Inconsistencies:** Sometimes DALL-E might produce inconsistent results for slightly varied prompts.
- **Sensitive to prompt structure:** The model can be finicky regarding how a prompt is phrased, meaning slight changes can yield vastly different results.

Bias and Representation

- **Reflection of cultural biases:** As with other AI models, DALL-E might inadvertently reproduce or amplify societal biases present in its training data.
- **Lack of diversity:** There are concerns about the potential underrepresentation or misrepresentation of certain groups, themes, or concepts.

Quality Control

- **Absence of artistic intuition:** While DALL-E can generate images, it doesn't possess artistic intuition. Some creations might be technically correct but lack an aesthetic appeal or emotional depth.
- **Boundary cases:** In certain edge cases, the generated images can be nonsensical or not aligned with the user's intent.

Dependency on Prompts

- **Vagueness:** A vague prompt might lead to an array of wildly varied outputs, making predictability a challenge.
- **Over-Reliance:** The over-dependence on prompts might limit spontaneous creativity, as the model always needs a textual nudge.

Lack of Contextual Understanding

- **Literal interpretations:** At times, DALL-E might interpret prompts too literally, missing nuances or metaphorical interpretations.
- **Missing broader context:** The model may not always capture the broader context behind certain prompts, especially if they require a deeper understanding or a multilayered interpretation.

Uncontrolled Outputs

- **Inappropriate content:** There's a possibility that DALL-E might generate content that's inappropriate or offensive, especially if it misinterprets a prompt.
- **Filtering challenges:** Implementing robust filters to prevent the generation of unsafe content can be complex.

Commercial and Economic Concerns

- **Monetization challenges:** As AI-generated art becomes more common, its value and the means to monetize it might become challenging topics.
- **Competition with traditional artists:** The ease and speed of generating content with DALL-E might pose challenges to traditional artists and designers in terms of pricing and value proposition.

Scalability and Access

- **Exclusivity:** The high computational costs might make advanced implementations of DALL-E exclusive to well-funded entities, thereby limiting broader access.
- **Dependency on OpenAI's infrastructure:** Individual users or developers are largely dependent on OpenAI's infrastructure and API access, which can influence the rate, cost, and scalability of applications.

While DALL-E represents a groundbreaking achievement in the realm of AI-driven creative processes, it's essential to approach its capabilities with a balanced perspective, acknowledging both its vast potential and the limitations and challenges it presents.

The Future of DALL-E

Expansion Beyond Static Images

- **Dynamic artwork and animations:** One of the most exciting prospects for DALL-E is its potential evolution from generating static images to producing animations or even short film-like sequences. By extending the model's capabilities, future iterations might be able to take a narrative prompt and generate a moving visual representation.
- **Interactive content creation:** As technologies advance, DALL-E could be integrated with virtual reality (VR) or augmented reality (AR) platforms, paving the way for interactive art or gaming experiences crafted in real time based on user interactions.

- **Integration with other sensory modalities:** The future could see DALL-E not just limited to visuals but integrated with other AI models to produce multisensory outputs, combining visuals with sound, touch, or even smell to create immersive experiences.

Collaborative Human-AI Artistic Endeavors

- **Co-creation platforms:** The next phase of DALL-E might focus on collaborative platforms where artists provide foundational ideas or sketches, and the AI refines, suggests, or expands on them, becoming a co-creator rather than just a tool.
- **Artistic training and learning:** Upcoming versions of DALL-E could be trained or fine-tuned with specific artistic styles or trends, offering personalized artistic assistants to creators. For instance, an artist could train DALL-E on their own creations, allowing the AI to generate content seamlessly in line with the artist's unique style.
- **Redefining artistic boundaries:** As AI becomes more integrated into the artistic process, it could challenge and expand the very definitions of art, creativity, and originality, leading to new genres and forms of expression.

Ethical and Societal Implications

- **Intellectual property and ownership:** As DALL-E's creations become more intricate and original, the lines between human- and AI-generated art could blur, necessitating a reevaluation of copyright laws and intellectual property rights. Who owns the rights to AI-generated art? The user who gave the prompt, the developers of the AI, or the AI itself?
- **Economic impacts on the art world:** As AI-driven artworks become more prevalent, they might impact the value and economic dynamics of the art market. Artists, galleries, and institutions will need to navigate this new terrain where AI-generated art might be sold alongside traditional works.
- **Bias and representation:** Like other AI models, future versions of DALL-E will need continuous monitoring and refining to ensure the artwork generated is free from biases and appropriately represents diverse cultures, identities, and perspectives.

Enhanced Personalization and Adaptability

- **Learning from user interactions:** Future iterations of DALL-E could be designed to learn and adapt based on user interactions, leading to highly personalized AI art assistants. For example, over time, DALL-E might recognize a user's preferences and proactively suggest artistic choices aligned with those tastes.
- **Adaptive environments:** In interactive platforms such as video games or virtual worlds, DALL-E could be used to adapt the environment in real time based on the player's actions, choices, or emotions, creating truly unique experiences for each user.

Increased Accessibility and Integration

- **Democratizing art creation:** By making powerful tools such as DALL-E more accessible to the public, individuals without traditional artistic training could express themselves visually, democratizing the realm of visual creation.
- **Cross-platform integration:** The future could see DALL-E being integrated into a variety of platforms, from graphic design software and movie production tools to educational platforms and beyond. This integration would streamline the content creation process and enhance the capabilities of existing platforms.
- **Open-source collaborations:** While DALL-E is a product of OpenAI, the principles behind its functioning could inspire open-source alternatives, leading to a rich ecosystem of AI-driven creative tools developed collaboratively by global communities.

In conclusion, the future of DALL-E promises a blend of technological advancements, artistic revolutions, and societal changes. While the opportunities are vast, it's essential for developers, artists, and society at large to navigate this future with a keen awareness of the ethical implications and challenges that such powerful tools present.

13

Text Synthesis with CTRL

Table of Contents

An Overview of CTRL

The burgeoning world of AI has seen several significant developments in recent years, one of which is Salesforce's Conditional Transformer Language Model, popularly known as CTRL. Unlike traditional language models that passively await a user's input to generate relevant responses, CTRL employs a more proactive approach. The uniqueness of CTRL lies in its use of control codes—a foundational element that steers its text generation capabilities.

Developed on the back of the transformer architecture, CTRL boasts a formidable 1.63 billion parameters, making it one of the most powerful language models during its introduction. However, what sets it apart isn't just its size but its ability to be conditioned based on certain parameters. These parameters, or control codes, are provided as input and give explicit instructions to the model about the kind of response desired by the user. For instance, one could specify a topic, tone, or even the source of information to guide the model's output.

The incorporation of control codes is a significant leap from traditional models. It imparts more user-driven directionality to AI-generated content. This ensures not just high-quality outputs but also ones that align closely with the user's intention. Whether it's drafting an email in a formal tone, generating news-style content, or even crafting fictional narratives, CTRL provides unprecedented customization in AI-driven text synthesis.

While CTRL was a breakthrough, it also illuminated the pathway for subsequent models and innovations. It demonstrated that the future of AI text generation wasn't just about creating larger models but smarter ones—models that can understand nuanced user requirements and adapt their responses accordingly. In essence, CTRL was more than just a milestone in the journey of AI; it represented a paradigm shift in how we perceive and harness the potential of machine-driven content generation.

In the expansive universe of AI language models, CTRL stands out as a pivotal innovation. Born from the transformer architecture, CTRL's distinctiveness is enshrined in its utilization of control codes, a mechanism that brings a tailored touch to the world of machine-driven text generation. Instead of operating on generalized algorithms, CTRL elevates customization by allowing users to employ specific conditions that influence the text it produces, resulting in outputs that are more relevant and in tune with the user's intent.

CTRL's design is underpinned by an impressive 1.63 billion parameters, granting it the computational power and flexibility to process vast swathes of information. However, it's the inclusion of control codes that marks a real departure from conventional models. These codes act as signposts, guiding the model in generating content by specifying elements such as topic, style, sentiment, and even the desired source of information. Think of control codes as the conductor of an orchestra, directing each section to produce a harmonized output.

The significance of CTRL extends beyond its immediate capabilities. By putting the power of directionality into users' hands, it redefines the relationship between humans and AI. It moves away from the paradigm where AI offers generic solutions and takes a step toward a more collaborative approach. Here, the AI doesn't just work *for* the user but works *with* the user, considering their specific needs and preferences.

But like any pioneering technology, CTRL also offers lessons for the future. It underscores the idea that the path forward for AI isn't merely in creating models with more parameters but in developing those that can understand and respond to nuances, adapting their knowledge base to produce content that isn't just accurate but also contextually relevant. By striking a balance between raw computational power and user-driven customization, CTRL sets the stage for the next chapter in AI, where machines not only comprehend our world but also cater to its myriad intricacies.

Prompts in CTRL: What's Different?

In the rapidly evolving realm of language models, the manner in which different models interact with prompts can greatly shape the utility and adaptability of the AI in real-world applications. CTRL presents a distinctive approach to this interplay compared to its predecessors and contemporaries. A dive into the nuances of CTRL's handling of prompts reveals a model striving for heightened contextual awareness and specificity, a feature that is quintessential for varied professional environments.

Control codes as advanced prompts: CTRL's standout feature is its implementation of control codes. Unlike conventional models that take a more general input as a prompt and produce an associated output, CTRL uses these control codes to determine the nature, tone, and context of the generated content. Where traditional models offer outputs based on broad training data and context, CTRL, with its control codes, allows users to dictate the theme, source, and even sentiment of the content. This represents a more direct form of interaction, where the model isn't just passively responding but actively tailoring its output to nuanced user needs.

Granularity of control: Previous language models operate on a principle of generalized intelligence, where the emphasis is on producing the most probable next word or phrase. CTRL, with its control codes, gives users the capability to achieve a finer granularity in the output. For instance, one can specify the style of writing (e.g. news headline vs. romantic novel), or even emulate specific sources (e.g. BBC vs. Wikipedia). This granularity transforms the model from a mere text generator to a versatile tool capable of fitting diverse applications.

Reduced ambiguity: A challenge with earlier models is the potential ambiguity of prompts. Given a generic input, the model might produce an array of outputs based on its vast training, leading to occasional mismatches with user intent. CTRL's control codes act as clarifying directives, drastically reducing ambiguities. By having codes that signify specific tones, styles, or sources, CTRL narrows down its potential outputs, aligning more closely with user expectations.

Learning from contextual missteps: It's worth noting that the journey of AI language models has been one of learning from limitations. Earlier models, at times, produced outputs that, while syntactically correct, were contextually amiss. CTRL, by integrating control codes, takes a significant leap in ensuring that its generated content is not just grammatically coherent but also contextually apt. This is a culmination of lessons learned from the contextual missteps of prior models.

Training and adaptability: While the underlying architecture and sheer size (in terms of parameters) play a pivotal role in the performance of language models, it's the training methodology that often dictates their real-world utility. Previous models, being more generalized, sometimes required iterative fine-tuning with prompts to achieve desired results. CTRL, on the other hand, is inherently designed to be more adaptable. The use of control codes means that users don't have to wrestle with the model to get what they want. Instead, by using the right codes, they can guide the model more effortlessly toward desired outputs.

In conclusion, CTRL's approach to prompts, embodied in its use of control codes, signals a shift in the language model paradigm. Instead of broad, generalized interactions, there's a move toward specificity, adaptability, and context awareness. While earlier models laid the groundwork by demonstrating the sheer potential of AI-driven text generation, CTRL builds upon this by offering a more refined, user-centric tool. It recognizes that in the vast sea of potential AI outputs, guiding the model to produce not just any content but the right content is paramount.

Showcasing CTRL in Action

CTRL is not just another addition to the ever-growing roster of language models. It brings about a refined mechanism of text synthesis, leveraging control codes to yield outputs of a specific nature, tone, and sentiment. To truly understand the capabilities of CTRL, one needs to see it in action across various scenarios. Let's traverse through some of these applications and witness firsthand the adaptability and prowess of this model.

News Article Generation

CTRL's ability to generate coherent and extensive text makes it an invaluable tool for journalism. By feeding it a control code specifying a news source style, say BBC, followed by a brief topic, the model can generate an article that reads as if it were written by a seasoned journalist. This not only aids in rapid content generation but also assists journalists in outlining articles or exploring different writing angles.

Literary Endeavors

CTRL is no less an artist than a tool. By using control codes for different literary styles or genres, such as "Victorian Novel" or "Modern Poetry," users can coax the model into crafting pieces that resonate with the chosen theme. For budding writers, this can serve as an inspiration fountain, offering them varied perspectives on a narrative or theme.

Business Communication

In a corporate setting, communication tone and clarity are pivotal. By employing CTRL with a control code such as "Formal business email," organizations can draft communications that maintain a professional standard. This becomes especially useful when drafting bulk communications or standardized responses, ensuring a consistent tone and style throughout.

Technical Documentation

CTRL's versatility also extends to generating technical content. By inputting a control code such as "Software documentation" and specifying a topic, users can get a well-structured documentation piece. While it may not replace human experts, it can surely assist them by offering a starting point or filling in standard documentation sections.

Multilingual Content Generation

With a vast understanding of multiple languages, CTRL can be a polyglot's dream tool. By providing a control code for a specific language and feeding in a topic or theme, the model can generate content in the chosen language. This becomes especially handy for organizations seeking rapid translations or multilingual content generation for diverse audiences.

Interactive Gaming

The gaming industry, especially the RPG (role-playing game) segment, can benefit immensely from CTRL. Game developers can utilize CTRL to craft game narratives, dialogues, or backstories. A control code specifying the genre or nature of the game (e.g. "Sci-fi game quest") can lead to generation of intricate plots or character interactions, enhancing the in-game experience.

Educational Content

In the realm of education, CTRL can act as a supplemental tutor. By using a control code such as "Historical overview" or "Math problem explanation," educators and students can access detailed explanations or summaries on a wide range of topics. This can aid in creating study materials or offering additional insights into complex subjects.

Legal Drafting

The legal profession, characterized by its specific jargon and format, can harness CTRL for drafting or reviewing documents. Inputting a control code such as "Contract clause" or "Legal brief" allows for the generation of content that adheres to legal standards, providing a valuable tool for paralegals and attorneys.

Entertainment Scripts

From movie scripts to podcast outlines, CTRL's prowess extends to the entertainment industry as well. By utilizing a control code such as "Romantic movie scene" or "Podcast intro," creators can get a generated piece that fits their content medium, serving as inspiration or a foundation for further refinement.

In the larger landscape of AI and language models, CTRL stands out not just because of its technical specifications but due to its real-world applicability. The model's structure, complemented by the ingenious implementation of control codes, opens up a realm of possibilities across diverse sectors. By showcasing its potential in varied scenarios, one can truly grasp the revolutionary capabilities of CTRL, reshaping the way we perceive AI-driven text synthesis.

Limitations and Ethical Concerns

Specificity of Control Codes

- CTRL operates on control codes to determine the type, style, or sentiment of the generated text.
- This requires users to have knowledge of the exact codes to use, potentially limiting its accessibility to the general public.
- There's also the risk of misinterpretation if the model doesn't recognize or incorrectly processes a control code.

Potential for Misinformation

- Like other language models, CTRL can generate content that is coherent but not necessarily accurate or factual.
- This raises concerns in areas such as news article generation, where a false or biased piece can mislead readers or propagate misinformation.

Over-Reliance in Professional Domains

- Although CTRL can produce content such as legal drafts or technical documentation, relying solely on it can introduce errors or overlook nuances.
- Human review and expertise remain crucial, and an over-reliance on the model can compromise the quality and accuracy of important documents.

Bias and Stereotyping

- Language models, including CTRL, are trained on vast datasets sourced from the Internet, inheriting the biases present in these datasets.

- Even with control codes, the model can occasionally produce outputs that reflect societal biases or perpetuate stereotypes, leading to ethical and PR challenges.

Lack of Emotional Understanding

- CTRL synthesizes text based on patterns learned during training but doesn't understand emotions or sentiments in the way humans do.
- This can lead to outputs that, while grammatically correct, might be perceived as insensitive or inappropriate in certain contexts.

Content Authenticity and Plagiarism Concerns

- With CTRL's ability to produce vast amounts of coherent text rapidly, there's potential for content to inadvertently resemble existing copyrighted material.
- This raises ethical concerns about authenticity, originality, and potential copyright violations.

Economic and Job Implications

- The adoption of CTRL in sectors such as journalism, legal drafting, or content creation can lead to concerns about job displacements.
- While it can act as an aid, there's a looming fear of it replacing human roles in certain domains, leading to economic and societal implications.

Data Privacy and User Trust

- Interactions with CTRL, especially in cloud-based implementations, can potentially expose sensitive user inputs to external entities.
- Ensuring data privacy, securing user data, and building trust are paramount, especially when the model is used in domains handling confidential information.

Over-Generation and Verbose Outputs

- CTRL, in certain scenarios, can produce outputs that are longer or more verbose than required.
- This can lead to information overload, making it cumbersome for users to sift through the content to find relevant details.

Dependency on Training Data

- The quality and diversity of CTRL's outputs largely depend on its training data.
- If trained on a narrow or biased dataset, the model's outputs can be skewed, limiting its adaptability and applicability across diverse topics and demographics.

Environmental Concerns

- Training advanced models such as CTRL demands significant computational resources, leading to environmental concerns due to the energy consumption of large-scale AI operations.
- Organizations need to weigh the environmental impact against the benefits offered by such models.

Ethical Usage and Misuse

- The power of CTRL can be harnessed for malicious intents, such as generating propaganda, false narratives, or misleading content.
- Ensuring ethical use and implementing safeguards against potential misuse are significant challenges faced by the AI community.

In summary, while CTRL offers a groundbreaking approach to text synthesis with its control codes, it doesn't come without challenges. Recognizing these limitations and ethical concerns is crucial for developers, users, and stakeholders. Only through a balanced understanding can we harness the full potential of CTRL while navigating its complexities responsibly.

CTRL's Evolution and Future Trajectories

From Simple to Complex—The Birth of CTRL

The journey of CTRL (Conditional Transformer Language Model) represents an evolution from simpler models to more complex ones, driven by the need for more control in text generation. Earlier models lacked the specificity and directive capacity to generate content tailored to specific themes, genres, or sentiments. CTRL emerged as a solution, introducing the concept of control codes. These codes were designed to allow users to guide the narrative, thus giving them more control over the output. As AI models continuously learned and expanded their knowledge base, CTRL became a testament to how text synthesis models were moving toward a more user-driven approach.

Advancements in Control Mechanisms

The essence of CTRL lies in its control codes. In its evolution, the emphasis has been on refining and expanding these codes for enhanced precision in output. Given the feedback from initial users and the AI research community, there has been a push toward developing more intuitive and diverse control codes. Future trajectories might see these codes becoming even more granular, allowing for micro-level control, or potentially the development of adaptive codes that learn and modify themselves based on user interactions over time.

Integration with Multimodal Models

The future of AI is not just about text—it's about integrating various data forms, from images to sounds. As CTRL evolves, there's potential for its integration with multimodal models. This could mean using text prompts to generate or modify visual or auditory content, leading to a more holistic creative process. For example, describing a scene using specific control codes in CTRL could lead to the generation of a corresponding image or video clip, signifying the convergence of textual and visual AI technologies.

Greater Personalization and Context Awareness

While CTRL offers more control through its predefined codes, the future may lie in models that can adapt to individual users or understand broader contextual nuances. Imagine a version of CTRL that understands the historical context of a piece of writing or can tailor its outputs to individual user preferences, learning over time. Such adaptive learning would not only make the model more user-friendly but also more effective in generating content that resonates with specific audiences or cultural contexts.

Ethical and Responsible AI Development

As with all AI advancements, the evolution of CTRL will necessitate greater emphasis on ethics. Given its power to generate diverse content, the potential for misuse or the propagation of misinformation is real. Future developments in CTRL might incorporate more robust mechanisms to identify and mitigate biases, ensure content authenticity, and promote ethical use. This could involve a combination of model training practices, user guidelines, and perhaps even in-built checks that warn or restrict certain types of content generation.

In conclusion, CTRL stands at a pivotal juncture in the landscape of AI-driven text synthesis. Its evolution from earlier models represents the broader trajectory of AI—moving toward greater user control, precision, and adaptability. As we look toward the future, CTRL might not just be a tool for generating text but could become an integral part of a more interconnected AI ecosystem, blending text with visuals, sound, and more. However, as with all advancements, it will be imperative to balance innovation with responsibility, ensuring that the power of such tools is harnessed for positive impact while mitigating potential pitfalls.

14

Learning Languages with T2T (Tensor2Tensor)

Table of Contents

Introduction to T2T

Tensor2Tensor, commonly referred to as T2T, represents a significant step in the evolution of machine learning libraries. Developed by the researchers at Google Brain, it's a flexible and extensible system that caters to the needs of the ever-evolving deep learning community. At its core, T2T aims to streamline the process of creating deep learning models, providing a platform that's both efficient and adaptive.

The initial inspiration for T2T stemmed from the realization that while there were numerous machine learning frameworks available, many lacked the flexibility and scalability required by serious researchers and developers. Traditional machine learning libraries tend to compartmentalize functions, treating data loading, preprocessing, model building, and training as distinct phases. However, in the dynamic world of deep learning, where rapid iteration and experimentation are key, this segmented approach can be cumbersome.

Enter Tensor2Tensor. Its name is suggestive of its primary function transforming one tensor into another. In computational terms, a tensor is a multidimensional array, and deep learning often involves translating input tensors (like images or sentences) into output tensors (like classifications or translations). T2T's framework is built around this fundamental idea of tensor transformations, making it especially suited for a vast array of machine learning tasks.

One of the areas where T2T has made a remarkable impact is in the domain of natural language processing (NLP). With language being an inherently complex and multifaceted medium, modeling linguistic phenomena requires a system capable of handling intricate tensor transformations. T2T provides a toolbox of predefined models and datasets, enabling researchers to jump-start their NLP projects. Instead of building models from scratch, they can leverage T2T's resources, adjusting and refining as necessary. This accelerates the model development process, allowing for quicker experimentation and, ultimately, innovation.

However, T2T isn't just limited to NLP. Its versatility extends to other areas of deep learning, including computer vision, speech recognition, and even game playing. It offers a unified interface for these diverse tasks, simplifying the user experience without compromising on functionality.

Another major advantage of T2T is its seamless integration with TensorFlow, Google's open-source machine learning framework. TensorFlow provides the computational backbone, handling intricate mathematical operations and ensuring efficient execution. T2T, on the other hand, acts as a user-friendly interface, abstracting away many of the complexities associated with deep learning model development. This synergy between TensorFlow and T2T ensures that users get the best of both worlds, computational power and ease of use.

In conclusion, Tensor2Tensor represents a transformative shift in the world of machine learning software. By prioritizing flexibility, scalability, and user experience, it has bridged the gap between advanced research and practical application. Whether you're a seasoned researcher looking to push the boundaries of what's possible or a novice hoping to dip your toes into the vast ocean of deep learning, T2T offers a comprehensive and intuitive platform to realize your ambitions.

The Role of Prompts in T2T

Foundational Framework

- T2T is a versatile machine learning library tailored for transforming one tensor into another, often utilized for language tasks.
- Prompts in T2T act as guiding tools or contextual frameworks that shape the output by creating a predefined expectation.

Contextual Guidance

- Prompts set the tone and context for neural networks to operate. By offering a starting point or a direction, they aid the model in generating desired outputs tailored to specific tasks or domains.

Integration with Seq2Seq Models

- Many tasks within T2T leverage sequence-to-sequence (Seq2Seq) models, which consist of encoder and decoder mechanisms.

- The encoder absorbs the input along with the prompt, crafting an intermediate representation. Following that, the decoder utilizes this representation to produce the expected outcome.

Influence of Attention Mechanisms

- T2T often employs attention mechanisms, which allow models to focus on particular segments of the input when generating corresponding parts of the output.
- With the integration of prompts, these mechanisms can give added weight to prompts, further reinforcing their guiding role.

Multitask Learning

- T2T's ability to perform multitask learning means a solitary model can simultaneously manage various tasks.
- Here, prompts play an instrumental role by indicating to the model which specific task it should concentrate on, resulting in varied tensor transformations based on the prompts provided.

Training Adaptability

- During the training phase, T2T models adapt to not only the primary dataset but also the format and semantics of prompts.
- This adaptability ensures that, over time, models can respond more effectively to broad or generalized prompts, increasing their flexibility in managing diverse input structures.

Zero-Shot and Few-Shot Learning

- In scenarios where models need to generalize from minimal data or address tasks they haven't been directly trained for, prompts prove invaluable.
- Through well-curated prompts, models can make better-educated guesses or inferences based on the knowledge they've accumulated.

Challenges and Precision

- The integration of prompts in T2T is not without challenges. The selected prompt can drastically alter model outputs, necessitating meticulous design and testing.
- Over-dependency on prompts might result in models becoming too inflexible or unable to generalize well without explicit prompt cues.

Feedback Loop and Dynamic Adaptation

- Feedback loops from users or systems can refine prompt effectiveness over time.
- There's potential in developing dynamic prompts within T2T that can adjust or evolve in real time based on instant feedback or changing needs, promising more accurate model guidance.

Evolving Research and Future of Prompts

- As the landscape of machine learning and T2T progresses, the role, integration, and optimization of prompts will continue to be central research areas.
- The exploration of prompts that can cater to more diverse tasks, or that can intuitively understand user needs, may revolutionize how T2T models are trained and applied.

In summary, prompts in T2T are not mere additions but critical components that deeply influence how models interpret and transform data. Through careful design, integration, and optimization, prompts can dramatically boost the model's efficiency and applicability in a myriad of real-world scenarios.

Practical Applications of T2T

Tensor2Tensor (T2T) is a library developed with the intention of managing tasks that involve converting one tensor (essentially multidimensional data arrays) into another. While this transformation might sound abstract, it's a cornerstone for a slew of crucial machine learning applications. Over time, T2T has become a valuable tool for a variety of language-related tasks due to its flexibility and efficiency. Here's a closer look at some practical applications where T2T has showcased its capabilities.

Machine Translation

One of the most lauded achievements using T2T has been in the domain of machine translation. The Tensor2Tensor models have been employed to build powerful systems capable of translating between different languages. Given the tensor-based nature of textual data, the capacity to transform sequences from one language tensor to another allows for dynamic translation. The success of such systems has facilitated real-time translations across numerous languages, breaking down communication barriers across the globe.

Text Summarization

The task of capturing the essence of lengthy textual content and distilling it into concise summaries has been facilitated through T2T. By understanding the content's tensor representation, models trained with T2T can generate a shorter sequence that encapsulates the original's primary meaning. This application is beneficial for news agencies, research organizations, and many other sectors that require concise renditions of extensive materials.

Image Captioning

While primarily intended for text, T2T has also shown promise in image-to-text conversions. Image captioning is about converting the tensor representation of an image into a textual description. This T2T application has seen extensive use in helping visually impaired users understand image content and also in cataloging vast amounts of visual data for organizations.

Speech Recognition

Voice-based interfaces and services are growing in popularity. Transforming audio tensors into textual tensors allows T2T models to convert spoken words into written text. This capability underpins voice assistants, transcription services, and various voice-activated technologies.

Sentiment Analysis

Companies are always on the hunt to gauge public sentiment about their products or services. T2T facilitates the transformation of textual data from social media, reviews, and comments into tensors representing sentiment, whether positive, negative, or neutral. Such insights can drive business strategies, product improvements, and marketing campaigns.

Autoregressive Tasks

For tasks where predictions are based on previous sequences, such as time series forecasting or even predicting the next word in a sentence, T2T's tensor transformation shines. These models can absorb past data tensors and project future trends or sequences, instrumental for financial forecasting, stock predictions, or next-word suggestions in typing interfaces.

Multitask Learning

One of the innovative uses of T2T is training models that can handle multiple tasks simultaneously. For instance, a single model might be trained for both translation and summarization. Here, prompts or specific inputs guide the model toward the task it should prioritize. This simultaneous multitask capability ensures efficient resource utilization and rapid deployment.

Custom Tasks

Given T2T's versatility, researchers and developers aren't restricted to predefined tasks. The library is designed to allow users to define their tensor-to-tensor transformations, paving the way for novel applications tailored to unique challenges, be it in health care, finance, or any other domain.

Reinforcement Learning

By using T2T, agents can be trained to interact with environments, receive feedback, and adjust their actions to maximize rewards. This is achieved by converting environmental states and agent actions into tensor forms, enabling sophisticated decision-making processes for games, simulations, and robotics.

Generative Tasks

Beyond its analytical capabilities, T2T also supports generative tasks. By feeding specific tensors, models can be prompted to produce original content, be it textual, visual, or audio-based. This aspect has exciting implications for creative fields such as art, music, and literature.

In sum, Tensor2Tensor is not just a tool but a versatile platform that has catalyzed advancements across a multitude of domains. Its tensor transformation ability, coupled with a robust architecture, ensures that it remains at the forefront of solving complex, real-world challenges in innovative ways.

Strengths and Weaknesses of T2T

T2T has ushered in a paradigm shift in handling deep learning and machine learning tasks, especially in the domain of NLP. Given its comprehensive nature and modular design, T2T has received both accolades and criticisms. Here, we'll delve into both the strengths and weaknesses of this notable framework.

Strengths

Versatility One of the most compelling features of T2T is its versatility. The framework is designed to accommodate a variety of machine learning tasks, ranging from machine translation to image recognition. This adaptability ensures that developers and researchers can use a unified platform for multiple projects, fostering consistency.

Efficient Training T2T has been optimized for efficient training. It provides high-speed training capabilities without compromising on the model's accuracy. This rapid training process is particularly beneficial for large-scale applications and when computational resources are at a premium.

End-to-End Approach T2T offers an end-to-end framework. Right from data preprocessing, model design, training, to the final inference, T2T provides tools and functionalities to manage every step. This holistic approach simplifies the development life cycle.

State-of-the-Art Models The framework is prepackaged with some of the state-of-the-art models in various domains, allowing developers to implement cutting-edge solutions without reinventing the wheel.

Flexibility in Model Design While T2T comes equipped with predefined architectures, it doesn't confine researchers to these alone. Its modular structure allows for custom model designs, encouraging experimentation and innovation.

Large Community Support Given its association with TensorFlow and its prominence, T2T enjoys robust community support. This backing ensures regular updates, a plethora of community-driven extensions, and a vast repository of knowledge for troubleshooting and best practices.

Weaknesses

Steep Learning Curve Despite its capabilities, T2T can be daunting for beginners. The abundance of features, combined with its modular architecture, might overwhelm newcomers. A solid understanding of TensorFlow and deep learning concepts is highly recommended.

Resource-Intensive T2T's models, especially the state-of-the-art ones, can be resource-hungry. Training these models requires significant computational power, which might be a deterrent for individual researchers or small-scale projects.

Overhead for Simple Tasks While T2T is brilliant for complex, multifaceted projects, it might be an overkill for simpler tasks. For basic applications, the overhead introduced by its comprehensive framework can be counterproductive.

Interdependency on TensorFlow T2T's deep integration with TensorFlow is a double-edged sword. On the one hand, it leverages TensorFlow's capabilities to the fullest. On the other, any major changes or shifts in TensorFlow can affect T2T users, making it harder to migrate to other platforms if needed.

Potential for Overfitting With its vast array of tools and functionalities, there's a potential risk of overfitting if not handled correctly. While the tools are powerful, they necessitate a deep understanding to ensure models generalize well to unseen data.

Updates and Deprecated Functions Given the rapid evolution of machine learning and TensorFlow itself, T2T undergoes frequent updates. These updates can sometimes deprecate older functions or introduce changes that might disrupt existing projects.

In conclusion, Tensor2Tensor is undeniably a powerful tool that has significantly influenced the machine learning landscape. Its strengths lie in its versatility, efficiency, and comprehensive nature, making it a preferred choice for many complex projects. However, it's essential to understand its limitations and the challenges it presents. Like any tool, its effectiveness is determined not just by its inherent capabilities but also by the proficiency of those wielding it.

The Future of T2T

T2T emerged as a transformative tool in the landscape of machine learning and natural language processing, integrating a diverse range of models and tools under a unified framework. The dynamic nature of AI and deep learning ensures that platforms such as T2T continue to evolve, both in response to technological advancements and the shifting needs of the community. As we look toward the horizon, several potential trajectories and developments come into view regarding the future of T2T.

Integration with Emerging Technologies

The future of T2T lies in its ability to seamlessly integrate with new and emerging technologies. Quantum computing, neuromorphic hardware, and advanced neural architectures are all frontier areas that could influence T2T's developmental trajectory. By being adaptable to these novel platforms, T2T could maintain its position at the forefront of the deep learning ecosystem.

Enhanced Modularity and Customizability

While T2T is known for its modular architecture, there's room for growth in terms of customizability. Developers and researchers are likely to demand more granular control over their workflows. Future iterations of T2T could offer more plug-and-play components, allowing users to effortlessly swap out parts of their pipeline to suit their specific needs.

Focus on Efficiency and Scalability

As models become more sophisticated, they also demand more computational resources. To counteract this, there will be an emphasis on optimizing T2T for more efficient training and inference. Techniques such as model distillation, pruning, and quantization might become integral components of the T2T framework, ensuring that state-of-the-art models remain accessible to a wide audience.

Strengthening Transfer Learning Capabilities

Transfer learning has shown immense promise in utilizing knowledge from one domain to aid performance in another. T2T's future may see it harnessing more advanced transfer learning methods, enabling models to generalize better across diverse tasks with limited data.

Ethical AI and Fairness Tools

The broader AI community is becoming increasingly conscious of ethical considerations. The future of T2T will likely involve integrating tools that assist in ensuring fairness, transparency, and ethical considerations in AI models. By building in functionalities that help detect and mitigate biases or offer explanations for model decisions, T2T could foster more responsible AI development.

Collaboration and Community Engagement

T2T's strength lies not just in its technical capabilities but also in its vibrant community. We can anticipate more collaborative tools within the platform, allowing researchers globally to cooperate on projects, share insights, or fine-tune models collectively. Such collaborative features would amplify the pace of innovation.

Diversification of Application Domains

While T2T has made significant inroads in natural language processing and computer vision, there's a universe of problems waiting to be addressed. From bioinformatics and climate modeling to financial forecasting and beyond, T2T might branch out, offering specialized tools and architectures for a broader range of domains.

Interoperability with Other Frameworks

The AI ecosystem thrives on diversity. While TensorFlow serves as the backbone for T2T, the future might witness T2T enhancing its interoperability with other prominent frameworks such as PyTorch, JAX, or MXNet. Such compatibility would allow for a richer exchange of ideas and methods across platforms.

Embracing the Edge

With edge computing gaining momentum, there's a push to run sophisticated AI models on edge devices. The future of T2T could emphasize tools and methods tailored for edge deployment, ensuring that powerful AI capabilities are available even on resource-constrained devices.

Comprehensive Learning Paradigms

Beyond supervised and unsupervised learning, the landscape of learning paradigms is vast. From self-supervised methods to reinforcement learning, T2T's future iterations might incorporate a broader spectrum of learning methodologies, offering researchers a comprehensive tool kit.

In closing, the potential trajectories for T2T are vast and diverse. As it continues to evolve, it embodies the very essence of the dynamic AI field ever-adapting, ever-improving, and always pushing the boundaries of what's possible.

15

Building Blocks with BERT

Table of Contents

Getting to Know BERT

BERT, which stands for Bidirectional Encoder Representations from Transformers, marks a transformative shift in the domain of natural language processing (NLP) and has been a cornerstone for numerous state-of-the-art models developed after its inception. Before diving deep into BERT's specifics, it's crucial to recognize the evolving trajectory of NLP. The field has undergone significant transitions, moving from rule-based systems to machine learning algorithms, and eventually to the deep learning architectures of today. These evolutionary steps aimed to better capture the complexities and nuances of human language. BERT emerged as a watershed model in this progression, uniquely setting itself apart from its predecessors.

Historically, most NLP models treated text in a unidirectional manner. That is, they read text either from left to right or vice versa. This approach had an inherent limitation it didn't fully capture the context of a word as humans do by understanding both its preceding and succeeding words. BERT, on the other hand, introduced a bidirectional approach. By reading text both ways—from left to right and from right to left—BERT could understand the context surrounding each word, resulting in a richer representation of text.

This bidirectional training is achieved using transformers, a type of deep learning architecture introduced in a paper titled "Attention is All You Need" by Vaswani et al. Transformers allow models such as BERT to focus on different parts of a sentence, adjusting this focus as needed to grasp the context better. This "attention" mechanism lets BERT decide which words in a sentence are most relevant when trying to understand a particular word's meaning.

While the architectural distinctiveness of BERT is impressive, what truly propelled it to the limelight was its pretraining approach. Traditionally, NLP models were trained on specific tasks, such as translation or question-answering, using labeled data. However, labeled data is scarce and expensive to produce. BERT's genius lies in its two-step training process pretraining and fine-tuning. During pretraining, BERT learns by predicting missing words in a sentence, using vast amounts of unlabeled text. This masked language model approach allowed BERT to gain a generalized understanding of language. Once pretrained, BERT can then be fine-tuned on a smaller, task-specific dataset, making it adaptable to a wide array of NLP tasks with minimal task-specific training data.

The outcomes of this dual-step training were groundbreaking. BERT achieved state-of-the-art results on eleven NLP tasks upon its release, spanning from sentiment analysis to question-answering. Its success can be attributed to its deep understanding of context, something previous models struggled with. For instance, consider the word "bank" in the sentences "I sat by the river bank" and "I went to the bank to withdraw money." While the word "bank" remains the same, its meaning differs based on the surrounding words. BERT, with its bidirectional understanding and attention mechanism, can differentiate these nuances with high accuracy.

In essence, BERT revolutionized the field of NLP not just by introducing a new architecture but by shifting the paradigm on how models are trained and fine-tuned. Its pretraining approach democratized NLP, enabling researchers and developers to achieve state-of-the-art results on specific tasks without needing vast amounts of labeled data. This democratization has led to an explosion of NLP applications, many of which are built on the foundational blocks provided by BERT. As the field moves forward, BERT's influence is evident, with many subsequent models borrowing from its principles while introducing their innovations. In understanding the history and trajectory of NLP, BERT undeniably stands out as a pivotal chapter.

How BERT Handles Prompts

Tokenization The first step in BERT's process is to convert the input text (prompts in this case) into tokens. These tokens can represent a whole word, part of a word, or even a single character, especially in languages where words can be broken down into smaller, meaningful units. BERT uses WordPiece tokenization, which breaks words into the most common sub-word units, allowing for a more efficient representation of a wide array of words, even those not seen during training.

Prepending Special Tokens Once the text is tokenized, BERT appends specific tokens to the input. The most common are [CLS] and [SEP]. The [CLS] (short for "classification") token is added to the start of the input, and it's the token whose final hidden state is used for classification tasks. The [SEP] (separator) token, on the other hand, is used to separate different segments of the text, ensuring BERT knows when one segment ends and another begins.

Embeddings Formation With the tokenized input ready, BERT converts each token into a high-dimensional vector using embeddings. BERT employs three types of embeddings:

- **Token embeddings:** These represent each token and capture the semantic meaning of the words or sub-words.
- **Segment embeddings:** For tasks that involve pairs of sentences (such as question-answering), BERT uses segment embeddings to differentiate between the two.
- **Positional embeddings:** Since transformers (the architecture BERT is based on) do not have a built-in sense of order or position, positional embeddings are added to provide the model with knowledge of the position of each token within a sequence.

The embeddings from these three sources are summed to produce a single vector for each token.

Processing through Transformer Layers BERT's core architecture consists of multiple layers of transformers. Each token's embedding is passed through these layers, where, using the attention mechanism, the model determines which parts of the text are relevant to understanding the context around a particular token. This context-aware processing ensures that even if a word appears multiple times in a sentence, its representation can vary depending on its surrounding words.

Masked Language Modeling One of the innovative training techniques BERT uses is the masked language model (MLM) approach. During training, some tokens in the input are randomly masked (hidden), and the model tries to predict them based on the surrounding context. This strategy forces BERT to learn a deep understanding of the context. However, when handling prompts, BERT doesn't predict masked tokens. Instead, it leverages the context understanding developed during the MLM training phase.

Generating Output Representations Once the embeddings pass through all transformer layers, BERT provides context-rich representations for each token. For classification tasks, the representation corresponding to the [CLS] token is generally used as it has aggregated information from the entire sequence. For token-level tasks such as named entity recognition (NER), the individual token representations are used.

Task-Specific Heads BERT is designed to be a base model that can be fine-tuned for various tasks. After obtaining the final contextual representations, they are passed to task-specific heads. For instance, for classification, a simple feed-forward neural network might be added on top of the [CLS] token's representation. For token-level tasks, heads might be designed to generate predictions for each token in the sequence.

Fine-tuning: While BERT is pretrained on a massive corpus, it is often fine-tuned on a smaller, task-specific dataset. During this process, the prompts play a crucial role. They provide the specific context and examples that allow BERT to adapt its generalized language understanding to the nuances of a particular task.

In summary, when handling prompts, BERT uses a series of intricate steps to ensure it captures the depth and breadth of linguistic context. Its approach to tokenization, embeddings, transformers, and fine-tuning allows it to offer state-of-the-art performances across a wide range of NLP tasks.

Use Cases and Real-World Examples

Sentiment Analysis

- Description: BERT can be fine-tuned to determine the sentiment of a given piece of text, whether it's positive, negative, or neutral. By understanding the context of words and the relationships between them, BERT excels in capturing the nuanced emotions in text.
- Real-world example: E-commerce platforms often employ BERT-based models to analyze product reviews. By gauging sentiment, companies can quickly identify issues, improve products, or highlight top-rated items.

Named Entity Recognition

- Description: NER involves identifying and classifying named entities in text into predefined categories such as names of persons, organizations, or dates. BERT's deep understanding of context helps in differentiating between ambiguous terms.
- Real-world example: News agencies use BERT for NER to automatically tag and categorize content, enabling efficient content management and retrieval.

Question-Answering Systems

- Description: BERT can be used to build systems that read a passage and answer questions related to it. It can find exact answers and is often used in combination with other models for more sophisticated systems.
- Real-world example: Google's search engine has integrated BERT to improve the matching of user queries with more relevant search results, especially for longer, more conversational queries.

Text Classification

- Description: This involves categorizing text into predefined groups. With its capacity to understand context and nuance, BERT has been fine-tuned for various classification tasks.
- Real-world example: Financial institutions deploy BERT models to classify news articles or financial reports, determining if they contain information that might influence stock prices.

Language Translation

- Description: While BERT isn't a sequence-to-sequence model (like some other models explicitly designed for translation), it can be used in multilingual settings and combined with other models to improve translation tasks.
- Real-world example: Multinational corporations use BERT to assist in translating content, ensuring that context and sentiment remain consistent across languages.

Semantic Text Similarity

- Description: BERT can be employed to determine how similar two pieces of text are in terms of meaning, making it valuable for tasks such as content de-duplication or plagiarism detection.
- Real-world example: Academic institutions utilize BERT-based systems to detect plagiarized content in research papers or assignments by gauging the semantic similarity to existing works.

Content Recommendation

- Description: By understanding the semantic content of texts, BERT can be utilized in recommendation systems to suggest articles, products, or media.
- Real-world example: Online news platforms employ BERT to recommend related news articles to readers based on the context of the articles they're currently reading.

Search Optimization

- Description: Traditional keyword-based search can be limited. BERT allows for more natural language–based search, understanding the intent behind queries.
- Real-world example: E-commerce platforms have integrated BERT into their search functionalities, allowing users to search with more conversational or vague queries and still get relevant results.

Chatbots and Conversational Agents

- Description: BERT can be integrated into chatbot systems to improve understanding and generate more contextually relevant responses.
- Real-world example: Customer support chatbots in tech companies have started to use BERT to better understand user issues and provide more accurate solutions.

Health Care and Medical Analysis

- Description: BERT can be fine-tuned on medical datasets to assist in tasks such as symptom checking, medical literature analysis, or drug discovery.

- Real-world example: Some hospitals have piloted BERT-based systems to analyze patient records, assisting doctors in diagnosis by drawing attention to critical information.

BERT's design, which emphasizes context understanding, has positioned it as a powerhouse in the realm of NLP. These use cases represent just the tip of the iceberg, and as more specialized versions of BERT emerge (such as BioBERT for biomedical texts), its applicability continues to expand across industries and domains.

Limitations of BERT

BERT has been a groundbreaking model in the world of NLP, delivering state-of-the-art results across a myriad of tasks. However, despite its impressive capabilities, BERT is not without its limitations.

Computational Intensity BERT's architecture, especially the larger versions such as BERT-Large, requires substantial computational power. Fine-tuning and training the model mandates high-memory GPUs, making it less accessible for individuals or institutions with limited resources. This computational demand also means that real-time applications can be challenging unless optimized versions or hardware acceleration are used.

Memory Footprint With millions of parameters, BERT models have a large memory footprint. This can be a challenge for deployment in resource-constrained environments, such as mobile devices or edge devices. As a result, there have been efforts to produce smaller, distilled versions of BERT that retain much of its power but at a fraction of the size.

Overfitting on Smaller Datasets Due to its large number of parameters, BERT can overfit when fine-tuned on smaller datasets. Careful regularization and hyperparameter tuning become essential in such scenarios. In some cases, using smaller versions of BERT or other regularization techniques may be more appropriate for limited data.

Lack of Transparency BERT, like many deep learning models, is often considered a black box. It can be challenging to interpret why BERT makes specific predictions or how it derives particular embeddings. This lack of transparency can be an issue in critical applications such as health care or finance, where interpretability and understanding model decisions are crucial.

Bias and Ethical Concerns BERT is trained on vast amounts of Internet text, which means it can inherit and even amplify biases present in that data. The model may, unintentionally, exhibit gender, racial, or other biases in its predictions, leading to ethical concerns, especially in sensitive applications.

Token Limitation BERT has a maximum token limitation (e.g. 512 tokens for BERT-Base and BERT-Large). This constraint means that longer texts have to be truncated, split, or otherwise managed, potentially losing vital information in processes such as document classification or information extraction from extensive documents.

Dependency on Context BERT's strength in understanding context can sometimes be its weakness. For tasks where the focus is on individual words or tokens rather than their interrelationships, BERT might be overkill, or it might miss nuances understood better by models specifically designed for such tasks.

Training from Scratch Challenges While BERT excels in fine-tuning tasks, training it from scratch requires vast amounts of data, computational resources, and expertise. This makes the original training of models such as BERT mostly confined to well-resourced organizations or research institutions.

Generalization While BERT achieves state-of-the-art results on many benchmark datasets, there's no guarantee it will always generalize well to specific real-world, niche tasks, or unseen data distributions. Tailoring and customizing the model might be required for specialized applications.

Language Limitations The original BERT model was trained on English text. Although multilingual versions exist, for many languages, especially those less represented on the Internet, the richness of embeddings and understanding might not be as refined as it is for English.

Adversarial Vulnerability Recent studies have shown that BERT, like many deep learning models, can be susceptible to adversarial attacks. Tiny, human-imperceptible alterations to input text can sometimes lead to drastically different model predictions, raising concerns in security-sensitive applications.

In conclusion, while BERT has undeniably revolutionized the landscape of NLP and has become a cornerstone in many applications, it's essential to be aware of its limitations. Understanding these constraints helps in making informed decisions about when and how to use BERT and how to navigate its challenges effectively.

Future Developments in BERT

BERT has indisputably carved a significant niche in the realm of NLP since its introduction. As with any technological marvel, the evolutionary trajectory of BERT is bound to be influenced by the challenges it currently faces, the emerging needs of the industry, and advancements in complementary technologies. Here's a glimpse into potential future developments in BERT.

Optimization for Efficiency As computational demands rise and the push for real-time processing becomes more prevalent, we can anticipate more efficient versions of BERT. These may feature fewer parameters without compromising on performance, using techniques such as model distillation, pruning, or quantization. The goal would be to make BERT more accessible and deployable in resource-constrained environments, from mobile devices to edge computing setups.

Improving Transparency and Interpretability One of BERT's criticisms is its black box nature. As the demand for model interpretability grows, especially in sectors such as health care or finance, future iterations or tools built around BERT might offer better insights into the model's decision-making processes. Techniques such as attention visualization, feature importance ranking, and model explanation methods might become more integrated with BERT architectures.

Addressing Biases Ethical AI and responsible machine learning are paramount. Future versions of BERT or its training methodologies will likely focus more on identifying, quantifying, and mitigating biases in predictions. This would entail both refining training data and post hoc techniques to ensure fairness and reduce inadvertent perpetuation of stereotypes.

Task-Specific Variants While BERT is designed as a general-purpose model, the future might see more variants tailored for specific tasks or industries. For instance, BERT models fine-tuned and optimized for legal, medical, or scientific jargon could emerge, catering to niche but essential sectors.

Cross-Lingual Advancements Multilingual BERT is just the beginning. The increasing globalization of the digital space necessitates models that can not only understand multiple languages but can also translate, transliterate, and bridge cultural nuances. We can anticipate more sophisticated multilingual versions of BERT or models that can seamlessly transition between languages within a single document.

Incremental Learning and Adaptation Currently, fine-tuning BERT on new data involves a static process where the model is adjusted to a new task. The future might see versions of BERT that can incrementally learn from new data streams, continuously adapting and refining their knowledge without forgetting previous learnings—a step toward more dynamic, lifelong learning systems.

Integration with Other Modalities NLP doesn't exist in a vacuum. As AI progresses, there's a push toward multimodal models that can understand and generate content across text, images, sound, and more. Future developments might see BERT being integrated with systems akin to DALL-E (for images) or Whisper (for audio), leading to more holistic AI systems.

Robustness and Security With the growing awareness of adversarial attacks in deep learning, future BERT models will likely emphasize being more robust against such threats. This would involve both architectural innovations and training methodologies that make the model more resilient against adversarial inputs.

Domain Adaptation Techniques Transfer learning, where knowledge from one domain is applied to another, is a strength of BERT. The future might see enhanced techniques for domain adaptation, allowing BERT to provide richer embeddings and better performance even when faced with data very different from its training corpus.

Better Integration with Downstream Tasks As BERT is primarily a feature extractor, efforts will be directed toward better integration with downstream tasks. This might involve architectures that allow more seamless flow of information between BERT's layers and the specific layers designed for the end task, optimizing performance and efficiency.

In wrapping up, the future of BERT is undeniably bright. The areas outlined above represent a mix of current challenges and visionary aspirations. With the relentless pace of advancement in AI and NLP, it's exciting to contemplate where BERT and its progeny might stand in the next few years. The continued evolution of this transformative model will undeniably influence and be influenced by the broader trends in AI research and applications.

16

Voice Synthesis with Tacotron

Table of Contents

Introduction to Tacotron

In the domain of voice synthesis, Tacotron stands as a paradigm shift. As a neural text-to-speech system developed by Google, Tacotron marries the intricacies of humanlike voice simulation with the power of deep learning, breaking down the barriers that once defined the limits of computer-generated speech.

Historically, text-to-speech (TTS) systems were segmented into disparate parts: text analysis and waveform generation. The former would dissect the provided text into phonetic components, while the latter would take these components to generate the corresponding sound waveforms. The process, while effective, often led to voice outputs that, despite being clear, lacked the naturalness and intonation of human speech. This lack of fluidity stemmed from the challenges of converting textual phonetic representations into nuanced vocal outputs. Each module had its inherent limitations, and their amalgamation into a cohesive voice often bore the unmistakable signature of machine generation.

Tacotron changed this narrative. Instead of relying on a multistage process, it introduced an end-to-end approach. The model takes a sequence of linguistic features derived from the input text and directly produces a corresponding sequence of spectrogram frames, which can be converted to audio. In essence, Tacotron merges the traditionally separated stages of a TTS system into one cohesive process.

Under its architecture, Tacotron employs a sequence-to-sequence model, which is reminiscent of those used in machine translation. This model contains an encoder and a decoder, with attention mechanisms playing a pivotal role. As the encoder processes the input text sequence, the attention mechanism then helps the decoder focus on relevant parts of the encoded text when generating the spectrogram frames. This approach not only simplifies the TTS process but also leverages the power of attention mechanisms to produce more natural-sounding speech.

One of the remarkable achievements of Tacotron is its ability to handle challenging linguistic elements, such as intonation, stress, and rhythm, which are essential for speech's naturalness. It can modulate these elements based on the context, a nuance that earlier TTS systems often missed. The system can also be trained on various datasets, allowing for voice customization and the generation of speech in multiple languages and tones.

Tacotron's real magic lies in its adaptability and potential for improvement. Google, after introducing the original Tacotron, soon unveiled Tacotron 2. This successor combined the sequence-to-sequence model with WaveNet, a deep generative model of raw audio waveforms.

By integrating WaveNet into its architecture, Tacotron 2 was able to produce speech that was almost indistinguishable from human voice in certain contexts. This marriage of technologies showcased the profound potential lying at the intersection of different deep learning models.

In summary, Tacotron is more than just another TTS system. It embodies the evolution of voice synthesis, powered by deep learning. By transforming the traditional multistage approach of TTS into an end-to-end process, Tacotron has set a new benchmark in the synthesis of natural, humanlike speech. Its innovations not only open up new avenues in voice technology but also provide a glimpse into the future, where the line between human-generated and machine-generated speech might become increasingly blurred.

The Significance of Prompts in Tacotron

Direct Input for Speech Synthesis

Tacotron, at its core, is a text-to-speech (TTS) system. Prompts, which are essentially the textual inputs, act as the primary source of information for the system. The model converts this text, via its sophisticated architecture, into audible speech. The quality and clarity of these prompts directly influence the resulting speech's accuracy and naturalness.

Contextual Intonation and Emphasis

While traditional TTS systems might produce flat, monotonous outputs, Tacotron uses prompts to derive not just words but also their intended emphasis and tone. Through the nuances embedded in the text, Tacotron can generate speech with appropriate rises, falls, stresses, and pauses, mimicking natural human speech patterns.

Handling of Linguistic Challenges

Prompts often contain linguistic elements that are inherently challenging for TTS systems, such as homographs or words with multiple pronunciations based on context (e.g. "lead" as in "to guide" vs. "a type of metal"). Tacotron uses the broader context of the prompt to decipher the correct pronunciation in such scenarios.

Customization and Special Instructions

Prompts can be augmented with special instructions or metadata that guide Tacotron in generating a specific kind of speech output. For instance, a prompt might include cues about desired pitch, speed, or emotional tone, allowing users to customize the resulting audio to a high degree.

Multilingual and Dialectal Adaptations

Given a diverse dataset, Tacotron can be trained to recognize prompts in different languages or dialects. This capability means that the system's response to prompts can be as varied and globally encompassing as the data it's trained on. A prompt in English might produce standard American English speech, while another in Mandarin would result in accurate Mandarin speech, demonstrating Tacotron's versatility.

Interactive and Dynamic Responses

In interactive settings, where user inputs (prompts) might change or be extended based on ongoing interactions, Tacotron can produce dynamic speech outputs that evolve with the conversation. This makes Tacotron a valuable asset in real-time applications such as voice assistants or interactive storytelling.

Training and Fine-Tuning

Prompts play a crucial role in training Tacotron. Large datasets comprising diverse prompts help in fine-tuning the model. The variety ensures that Tacotron becomes proficient in handling a wide range of linguistic inputs, from simple everyday sentences to complex technical jargon.

Assessment of Model Quality

During the development and post-training phases, specific prompts are used to evaluate Tacotron's performance. These benchmark prompts help in assessing the system's accuracy, naturalness, and overall quality. Feedback from these evaluations can be looped back into refining the model.

Exploration of Creative Applications

Beyond straightforward speech synthesis, prompts can be used creatively with Tacotron. For instance, musical notes or rhythm-related instructions embedded in prompts can guide Tacotron in producing speech that aligns with a specific melody or rhythm, paving the way for innovative audio applications.

Error Handling and Feedback Loop

Not all prompts result in perfect speech synthesis. Some might expose weaknesses or blind spots in Tacotron's processing. These instances, while being challenges, are invaluable. By analyzing the discrepancies between the prompt and the undesired output, developers can iterate on the model, making it more robust and accurate in future interactions.

In essence, prompts serve as the vital bridge between human intent and Tacotron's speech synthesis capabilities. They not only provide the raw material for conversion but also offer context, emotion, emphasis, and customization options. The role of prompts, thus, is not just functional but also critical in advancing the state of the art in voice synthesis with Tacotron.

Showcasing Tacotron in Real-World Scenarios

Tacotron, as a leading voice synthesis system, has transcended the bounds of laboratory settings and found applicability in several real-world scenarios. Its ability to generate near-human, coherent, and contextually relevant speech has positioned it as a tool of choice for a myriad of industries and applications. Below, we delve into some of the practical scenarios where Tacotron has been deployed or holds promise.

Voice Assistants and Smart Home Devices

One of the most ubiquitous applications of Tacotron is in voice-activated assistants such as Google Assistant, Siri, and Alexa. These systems, embedded in smart devices, require clear, humanlike speech to interact with users. Tacotron provides them with the ability to respond to user prompts with natural-sounding answers, creating a more engaging and interactive experience.

Audiobooks and Reading Applications

The demand for audiobooks has surged in recent years. Tacotron can be employed to convert vast volumes of text into speech, eliminating the need for human narrators. Furthermore, its capacity to handle different tones and contexts ensures a pleasant listening experience. This capability also extends to applications assisting visually impaired individuals by reading out written content.

Interactive Voice Response (IVR) Systems

Businesses use IVR systems to guide callers through menu options or to provide automated information. Tacotron's proficiency ensures callers encounter a more humanlike interaction, enhancing user satisfaction and streamlining the customer service process.

Language-Learning Platforms

Platforms such as Duolingo or Rosetta Stone, which aim to teach new languages, can leverage Tacotron to produce accurate pronunciations and dialectal variations, providing learners with an authentic language acquisition experience.

Gaming and Virtual Reality

Immersive gaming experiences rely heavily on sound. Tacotron can be employed to generate dialogues for characters, especially in games that adapt to player choices, requiring dynamic voice responses. In virtual reality scenarios, it can provide voice overs or assist in interactive simulations.

Telecommunication

In scenarios where real-time translation is needed, Tacotron can be coupled with translation models to provide instant voice translation, bridging communication gaps in multilingual contexts.

Film and Animation Industry

Creating voice overs for animated characters or employing dubbed voices in movies is resource-intensive. Tacotron offers a solution, especially for preliminary versions or for animations with tight budget constraints. Its capacity to modulate voice tones can be harnessed to fit diverse characters.

Medical and Therapeutic Applications

In therapeutic contexts, Tacotron can be used to develop speech therapy tools. For patients recovering from strokes or speech impediments, software can be devised using Tacotron to assist in speech-training exercises. Additionally, in mental health scenarios, it can be used to create interactive therapeutic voice assistants.

Navigation and Automotive Industry

Modern vehicles come equipped with voice-guided navigation systems. Tacotron's synthesis can guide drivers with clear, context-aware instructions, enhancing road safety. In the realm of autonomous vehicles, it can interact with passengers, updating them about the journey or addressing their queries.

Research and Development

Beyond commercial applications, Tacotron serves as a pivotal tool in linguistic research, helping scholars in studying phonetics, tonality, and speech patterns. Its vast capabilities provide a practical way to experiment with and understand human speech nuances.

Custom Voice Alerts

In industrial settings or specialized applications, Tacotron can be employed to generate custom voice alerts or instructions, guiding workers or users in real time.

The widespread applicability of Tacotron underscores not just the advancements in voice synthesis technology but also the increasing human reliance on voice interfaces for diverse interactions. As the boundary between synthesized and human voice continues to blur, Tacotron stands as a testament to the remarkable strides made in the realm of artificial voice generation. Its incorporation across sectors and scenarios signifies the onset of an era where human-computer interaction is as natural as human-to-human dialogue.

Limitations and Room for Improvement

The advent of Tacotron and its successors in the realm of voice synthesis marked a significant step forward in generating humanlike speech. However, as with any pioneering technology, Tacotron is not without its limitations. These challenges highlight areas ripe for exploration and refinement in subsequent iterations or alternative models.

Consistency in Long Sentences

One of the inherent challenges Tacotron sometimes faces is maintaining voice consistency, especially in longer sentences. There might be slight fluctuations in tone or pace, which, though subtle, can differentiate synthesized speech from natural human speech.

Handling Complex Emotions

Human speech is replete with emotional nuances that convey more than just words. Whether it's the subtle tremble of anxiety or the warmth of joy, human voices are incredibly expressive. Tacotron, while impressive, isn't always adept at capturing these intricate emotional variations, particularly when they're blended or subtle.

Accent and Dialect Limitations

While Tacotron can be trained on various accents or dialects, its output is usually limited by its training data. For languages and dialects with limited datasets, the synthesized voice might lack authenticity. Moreover, switching between accents fluidly, as bilingual speakers often do, remains a challenging frontier.

Dependency on Quality Training Data

Tacotron's performance is intrinsically tied to the quality and diversity of its training data. Poor quality or biased training data can lead to suboptimal voice synthesis. This can result in mispronunciations or unnatural intonations.

Handling Uncommon Words and Sounds

Tacotron might stumble upon rare words, technical jargon, or non-standard sounds. These could be pronounced incorrectly or might be rendered with a generic tone, lacking the context-specific emphasis a human might naturally employ.

Computational Intensity

Real-time voice synthesis demands significant computational resources. While this might not be a hurdle for big tech companies, it can limit the deployment of Tacotron in low-resource environments or in smaller, portable devices.

Understanding Context

Sometimes, the meaning of a word or phrase depends on its context. Tacotron might occasionally misinterpret such contexts, leading to misplaced emphasis or incorrect intonation. For instance, the word "lead" can be pronounced differently based on whether it refers to the metal or the act of leading.

Robustness to Noisy Inputs

In real-world scenarios, voice synthesis models such as Tacotron might have to deal with noisy input data. The model's resilience to such inconsistencies and its ability to produce clear speech regardless is an area needing improvement.

Prosodic Elements

Elements such as rhythm, stress, and intonation—collectively known as prosody—play a pivotal role in human speech. While Tacotron has made strides in this domain, there's still room for enhancing its prosodic modeling to achieve a truly natural speech rhythm.

Fine-Tuning and Customization

Customizing Tacotron for specific voices, tones, or styles requires retraining on new datasets. Simplifying this customization process can democratize voice synthesis, enabling more creators to design unique voice experiences without extensive machine learning expertise.

Ethical Concerns

While not a technical limitation, the potential misuse of voice synthesis for creating deepfakes or impersonating voices raises ethical dilemmas. As Tacotron becomes more sophisticated, it's imperative to develop mechanisms to detect synthesized speech and ensure ethical usage.

Despite these challenges, the trajectory of Tacotron and voice synthesis, in general, remains promising. The ongoing research, coupled with an ever-expanding ecosystem of tools and datasets, indicates a future where these limitations will be progressively addressed. As we venture forward, the synergy of technology and human ingenuity will continue to refine the boundaries of what's possible, ushering in an era where synthetic voice becomes indistinguishable from the real thing.

The Next Steps for Tacotron

The journey of Tacotron in the field of voice synthesis represents an exciting blend of technological advancement and unbridled potential. While Tacotron and its subsequent iterations have made commendable strides, the road map ahead is packed with opportunities, challenges, and transformative prospects. Here's a discussion on the anticipated next steps for Tacotron.

Improving Emotional Resonance

A crucial frontier for Tacotron is to master the intricate art of infusing speech with nuanced emotions. Humans convey a myriad of feelings through subtle vocal shifts—be it the quiver of nervousness or the rhythm of excitement. The next versions of Tacotron will likely aim to capture and reproduce these emotional subtleties more proficiently, making synthesized speech virtually indistinguishable from a human's emotional expressions.

Expand Linguistic Diversity

Today's globalized world is a melting pot of languages and dialects. To cater to diverse audiences, Tacotron's future iterations will need to encompass an even broader spectrum of languages, dialects, and regional accents. This would entail not just training on diverse datasets but understanding and replicating the cultural and linguistic nuances associated with each.

Energy-Efficient Models

As edge computing becomes prevalent, there's a growing need for deploying advanced models such as Tacotron on devices with limited computational resources. Therefore, creating lightweight, energy-efficient versions without compromising on voice quality will be a focus area. This can open doors for Tacotron's integration into a broader array of devices, from wearable tech to low-power IoT devices.

Real-Time Adaptability

Future Tacotron models might be equipped with real-time adaptability features, enabling them to learn and adjust to a user's preferences or feedback instantly. This could be in terms of preferred tone, pace, or even nuances in pronunciation, resulting in a more personalized voice synthesis experience.

Integration with Multimodal Systems

As technology pivots toward multimodal interfaces (where interactions span across text, voice, visuals, and more), Tacotron's integration into such systems will be key. This means Tacotron would not only generate speech but also sync seamlessly with visuals, gestures, or other sensory feedback mechanisms, offering a holistic user experience.

Enhanced Prosody Modeling

While Tacotron has made inroads into modeling prosodic elements of speech (such as rhythm and intonation), there's still room for refinement. Achieving a natural-sounding speech rhythm, stress pattern, and intonation that fluidly adjusts to varying contexts will be an essential milestone.

Addressing Ethical Challenges

As voice synthesis technology becomes more sophisticated, its potential misuse—such as creating voice deepfakes—becomes a grave concern. An essential step for Tacotron and its community will be to embed mechanisms that can detect synthesized speech or watermark it, ensuring ethical and transparent usage.

Open-Source Collaborations

OpenAI's approach of collaborating with the broader research community has catalyzed innovations in the past. The next steps for Tacotron could involve more open-source initiatives, community-driven enhancements, and collaborative research projects to address its current limitations and explore uncharted territories.

Interdisciplinary Research

The future of Tacotron might witness a confluence of research from various domains—linguistics, neuroscience, psychology, and more. Such interdisciplinary collaborations can provide deeper insights into human speech patterns, emotions, and perceptions, refining Tacotron's capabilities further.

Enhanced Security Features

With voice becoming a pivotal mode of authentication in various applications, from banking to smart home devices, ensuring the synthesized speech is secure and tamper-proof will be paramount. Future Tacotron iterations might embed advanced encryption and security protocols, safeguarding the generated voice data.

In essence, while Tacotron has already redrawn the boundaries of voice synthesis, the horizon ahead is luminous with possibilities. Each challenge presents an opportunity for innovation, and each innovation brings us a step closer to realizing the dream of flawless, humanlike voice synthesis. The collaboration of researchers, developers, ethicists, and users will shape Tacotron's journey, ensuring it not only mimics human speech but resonates with the very essence of human communication.

17

Transformers in Music with MuseNet

Table of Contents

Introduction to MuseNet

In the rapidly evolving world of AI, OpenAI has continually sought to push the boundaries of what machines can achieve, particularly in creative tasks. MuseNet, one of their groundbreaking creations, stands as testament to this pursuit. As a deep learning model, Muse-Net leverages the power of transformers—specifically, the generalized transformer architecture—to generate music compositions spanning a range of styles and genres.

MuseNet's foundation rests on the transformer architecture, which has revolutionized natural language processing tasks with models such as GPT and BERT. However, MuseNet's ambition goes beyond words; it delves into the realm of melodies, rhythms, and harmonies. This ambitious crossover from text to music showcases the versatility of the transformer's design. With a massive 72-layer architecture and an equally impressive eight billion parameters, MuseNet digests vast amounts of musical data, ranging from classical works of Mozart and Beethoven to contemporary genres such as pop and rock.

What makes MuseNet particularly intriguing is its ability to merge different styles seamlessly. For instance, if a user provides a seed melody in the style of a Baroque composer and asks for a continuation in the style of a jazz musician, MuseNet can generate a coherent and harmonically rich composition that bridges the two distinct styles. This blending of styles isn't just a simple juxtaposition; the model genuinely understands the intricacies of each musical genre and crafts compositions that reflect a deep awareness of musical conventions and traditions.

Training MuseNet, like other transformer models, required vast datasets. OpenAI fed it with MIDI files—a digital sheet music format that encodes notes, rhythms, and instruments. The MIDI format, being more structured than raw audio, allowed MuseNet to focus on the essence of the music: the relationship between notes, the progression of harmonies, and the dynamics of rhythm. As it trained, the model began to understand not just how melodies progressed but also the subtleties that give different musical genres their unique identities.

The user interface for MuseNet offers a promising glimpse into the potential applications of AI in music. Users can select a composer or style, provide a seed melody or choose from a predefined list, and then watch as MuseNet crafts a unique musical piece based on those inputs. For educators, composers, and musicians, MuseNet serves as a powerful tool for inspiration, music analysis, and even education.

However, it's essential to acknowledge that while MuseNet's compositions are technically sound and often aesthetically pleasing, they don't necessarily carry the emotional depth or intentionality inherent in human compositions. MuseNet doesn't feel emotions or possess a lived experience, so its compositions are creations of patterns and learned data rather than genuine emotional expression.

In conclusion, MuseNet represents a significant stride in the world of AI-generated music. By harnessing the transformer architecture's power, it not only generates music across a wide array of styles but also showcases the immense potential that AI holds in the realm of creative arts. While it may not replace human composers or the emotional depth they bring to their creations, MuseNet undoubtedly opens doors to new possibilities, collaborations, and innovations in the ever-evolving landscape of music.

Role of Prompts in MuseNet

Inspiration for Musical Creation

At its core, MuseNet uses prompts to derive the initial inspiration for generating pieces of music. This means that users can provide a short melody, chord sequence, or even text-based instructions as a starting point. MuseNet takes this input and constructs a musical piece that follows or expands upon the initial prompt, ensuring the generated composition is harmoniously and stylistically consistent.

Guiding Genre and Style

MuseNet was trained on various genres of music, from classical to rock to jazz. Through prompts, users can specify a particular style or composer they want the generated music to emulate. For instance, a prompt such as "Compose a piece in the style of Mozart" would lead MuseNet to generate a piece reminiscent of Mozart's works, while "Produce a jazz improvisation" would yield an entirely different sound.

Dynamic Transitions and Mash-Ups

One of the unique features of MuseNet is its ability to transition between different styles seamlessly. Through carefully designed prompts, users can guide MuseNet to create mash-ups or compositions that transition from one genre or composer's style to another, offering a unique blend of musical flavors that might be challenging for human composers.

Lyric Integration

While MuseNet's primary strength lies in instrumental composition, prompts can also guide the model in accompanying a set of lyrics or generating melodies for existing lyrics. This means users can provide a textual prompt, perhaps a couple of verses, and MuseNet can suggest a potential melody that complements the words.

Emotional and Thematic Guidance

Prompts play a vital role in guiding the emotional or thematic tone of the generated music. For instance, a prompt suggesting "a melancholic rainy day" could lead MuseNet to produce a slow, introspective piece, while "a festive celebration" might yield an upbeat, lively composition. This feature is especially valuable for content creators who need background music tailored to specific scenes or moods.

Iterative Feedback Loop

MuseNet allows for an iterative process where users can refine their prompts based on the outputs they receive. If a generated piece is close but not quite right, users can adjust their prompts, add more specific instructions, or provide a different musical snippet to guide the model in a desired direction. This iterative feedback is crucial in achieving a final piece that aligns closely with the user's vision.

Expanding Short Melodies

Musicians or creators who have a brief melody in mind but are unsure how to expand it can use MuseNet to fill in the gaps. By inputting their short melody as a prompt, MuseNet can extrapolate and provide a full composition, offering a quick way to flesh out musical ideas.

Replicating Traditional Structure

If users are looking for compositions that follow traditional musical structures, such as sonatas or rondos, they can use prompts to instruct MuseNet accordingly. The model then generates music that adheres to the specific structure's rules, ensuring that the piece not only sounds good but also follows classical conventions.

In essence, prompts in MuseNet act as a bridge between the user's vision and the vast musical knowledge encoded within the model. They guide the model's creative process, ensuring outputs that are not only harmonious and pleasing but also align closely with the user's specific requirements and preferences. This dynamic interplay between human creativity and AI capability is what makes MuseNet a revolutionary tool in the realm of music composition.

Examples of MuseNet Outputs

MuseNet is a groundbreaking representation of the application of transformer models to the world of music. It creatively synthesizes compositions by merging different genres, instruments, and styles, ushering in a revolution in the realm of algorithmic music generation. Let's delve into some real-world examples of the output compositions MuseNet is capable of generating

Beethoven Meets The Beatles

Imagine the robust, dramatic notes of Beethoven's Fifth Symphony merging seamlessly with the iconic rock strains of The Beatles. MuseNet has been showcased to create such masterpieces, where the classical symphony transitions smoothly into a rendition reminiscent of "Hey Jude" or "Let It Be." This experiment not only manifests the model's understanding of individual musical identities but also its capability to fuse them without any jarring inconsistencies.

A New Take on Adele

Adele's "Rolling in the Deep" is instantly recognizable with its powerful vocals and rhythmic undertones. But what if it could be reimagined as a 1920s jazz piece? MuseNet takes on this challenge and transforms the pop anthem into a vintage jazz track. The powerful belting is replaced by sultry, smoky vocals, the modern percussions give way to the lively beats of jazz drums, and the entire composition is peppered with brass instruments, offering a completely fresh perspective on a familiar tune.

Crafting Cinematic Magic

Inputting the description of a dramatic, rain-soaked romantic reunion scene from a movie, MuseNet generated a poignant composition that wouldn't be out of place in a blockbuster movie score. Starting with gentle, cascading piano notes, symbolizing raindrops, the music swells into a full orchestral crescendo, capturing the tumultuous emotions of such a scene.

Bach Goes Metal

In a whimsical experiment, the melodies of Johann Sebastian Bach, characterized by their intricate patterns and baroque style, were seamlessly integrated with the aggressive and powerful world of heavy metal. The resulting piece was a harmonious blend of Bach's sophisticated harmonies and the raw energy of metal guitar riffs.

Cultural Journeys

MuseNet's prowess isn't limited to Western music. A prompt asking for a "Chinese guzheng-inspired tune with elements of reggae" resulted in a piece that began with the distinctive plucks of the guzheng, a traditional Chinese instrument, followed by the relaxed, off-beat rhythms characteristic of reggae. The fusion was not just sonically pleasing but also showcased the model's understanding of diverse musical elements.

Reimagining Pop Culture

The theme of *Game of Thrones* is universally recognizable. When prompted to reimagine it in the style of an 80s video game, MuseNet produced a chiptune version of the theme, reminiscent of the music from arcade classics. The epic orchestral anthem was transformed into a quirky, nostalgic tune, yet retaining the essence of the original.

The Sound of Literature

When provided with a Shakespearean sonnet, MuseNet was tasked to produce a piece that could serve as background music for a recitation. The output was a gentle Renaissance-inspired melody, replete with lutes and harpsichords, evoking the era in which Shakespeare wrote.

Holiday Festivities

A prompt describing "Christmas morning, with children opening presents under a snow-clad tree" led MuseNet to create a joyous composition dominated by bells, chimes, and light orchestral elements, capturing the essence of festive happiness and winter warmth.

Dancing with Tchaikovsky

When asked to produce a waltz reminiscent of Tchaikovsky's style, MuseNet delivered a piece that felt like a lost track from *The Nutcracker* replete with swirling violins and grand orchestral sweeps.

Futuristic Beats

A futuristic, cyberpunk-themed cityscape description prompted an electronic composition, with synthesized beats and robotic overtones, perfectly capturing a neon-lit, tech-driven future.

These examples, drawn from various demonstrations and experiments, underline MuseNet's deep-rooted understanding of myriad musical elements. Through MuseNet, OpenAI has heralded a future where AI doesn't just replicate existing art but actively contributes to creating new, previously unimagined artistic expressions.

Limitations and Ethical Considerations

MuseNet, with its capability to generate captivating musical compositions across a myriad of genres and styles, is an impressive application of transformer models in the realm of music. However, like all AI models, it comes with its own set of limitations and poses various ethical considerations.

Over-Reliance on Training Data

- MuseNet's outputs are heavily influenced by the training data it has been exposed to. This means it might replicate existing biases or inadvertently create compositions resembling copyrighted pieces.
- Ethical implication: Inadequate representation in the training set can lead to a lack of diversity in generated outputs or unintentional replication of problematic motifs.

Lack of Original Creativity

- MuseNet doesn't possess a creative spark or emotional intent. Its compositions are based on patterns and structures it has learned.
- Ethical implication: Overvaluing AI-generated music can undermine human creativity and the deep emotional and cultural resonance inherent in human-made compositions.

Copyright Concerns

- MuseNet can generate pieces that, unintentionally, might be very similar to existing copyrighted works, leading to potential legal conflicts.
- Ethical implication: Determining the ownership of AI-generated music remains a gray area. If a generated piece closely mirrors an existing work, it raises questions about intellectual property rights.

Commercial Exploitation

There's a potential for businesses to exploit MuseNet for mass-producing generic music without compensating human artists.

Ethical implication: This can devalue the labor and talent of human musicians, potentially pushing them out of commercial opportunities.

Loss of Cultural Nuance

- While MuseNet can mimic styles, it may lack the cultural, historical, or emotional depth tied to certain musical genres.
- Ethical implication: The commodification and superficial replication of culturally significant music can lead to appropriation concerns.

Quality Inconsistency

- Not all of MuseNet's outputs will be of high quality or make musical sense. Some might be repetitive or lack cohesiveness.
- Ethical implication: A surge of AI-generated music might lead to an influx of subpar compositions in the music ecosystem, affecting the overall quality of available music.

Data Privacy

- If MuseNet evolves to incorporate user feedback for improving outputs, there's a potential risk of user data being used without clear consent.
- Ethical implication: Without strict privacy controls, the line between enhancement through feedback and infringement on user privacy becomes blurred.

Job Displacement

- The more advanced MuseNet becomes, the more it poses a threat to certain sectors of the music industry, especially for composers of background scores, jingles, or other generic musical pieces.
- Ethical implication: Relying heavily on AI for music creation can lead to reduced opportunities for budding artists and professionals in the music industry.

Environmental Concerns

- Training and running powerful models such as MuseNet require substantial computational resources, leading to significant energy consumption.
- Ethical implication: The environmental footprint of advanced AI models needs to be considered, especially when used extensively for applications such as music generation.

Emotional Disconnect

- Music often serves as an emotional outlet for both creators and listeners. While MuseNet's compositions can be technically sound, they might lack the emotional depth and intent a human composer infuses into their work.
- Ethical implication: An over-reliance on AI-generated music might dilute the emotional richness of the musical landscape.

In summary, while MuseNet offers exciting possibilities in the domain of music generation, it's crucial for users, creators, and stakeholders to be aware of its limitations and the ethical implications tied to its widespread use. Balancing the potential of such technology with conscious considerations will determine its beneficial or detrimental impact on the music industry and society at large.

What Lies Ahead for MuseNet

As MuseNet continues to impress with its ability to produce diverse and intricate musical compositions, it is essential to look ahead and envisage its possible trajectories in the evolving landscape of AI and music. Here are key points that might define the future of MuseNet.

Integration into Music Production Tools

Forecast: MuseNet's capabilities could be seamlessly integrated into digital audio workstations (DAWs) and music production software.

Impact: This would allow music producers to utilize MuseNet for instant inspiration, generate base tracks, or even fill in gaps in their compositions.

Collaborative Compositions

Forecast: MuseNet might evolve to enable real-time collaborative features where it interacts with human musicians, providing suggestions, variations, or even entirely new sections of music based on real-time inputs.

Impact: A synergized man-machine collaboration could revolutionize how music is created, allowing for new genres and styles to emerge.

Improvements in Training and Versatility

Forecast: Future iterations of MuseNet may undergo training with even more diverse datasets, incorporating music from lesser-represented cultures, eras, and experimental genres.

Impact: The model's outputs would become even more diverse, breaking away from predominantly Western-centric compositions and potentially leading to a global music renaissance.

Personalized Music Creation

Forecast: Leveraging user feedback, preferences, and historical data, MuseNet might be able to tailor compositions to individual tastes.

Impact: Users could have personalized soundtracks, theme tunes, or even songs created on the fly, enhancing their musical experiences.

Ethical Guidelines and Copyright Framework

Forecast: As MuseNet's compositions get more recognition, there will be a push for clearer guidelines on copyright, ownership, and usage rights of AI-generated music.

Impact: The music industry might undergo legal and structural changes to accommodate and regulate AI-generated content, ensuring fair use and remuneration.

Expansion into Interactive Media

Forecast: MuseNet could find applications in video games, virtual reality, and augmented reality, producing adaptive music that responds to user actions and environments.

Impact: This would enhance immersive experiences in interactive media, making them more dynamic and responsive.

Educational Applications

Forecast: MuseNet might be used as an educational tool, helping students understand music theory, composition techniques, and the evolution of various musical styles.

Impact: Music education would become more interactive and personalized, potentially nurturing a new generation of composers and musicians.

Open-Sourcing and Community-Driven Enhancements

Forecast: There could be a move toward making MuseNet's architecture or parts of its dataset open-source, allowing the global developer community to contribute improvements.

Impact: Community-driven enhancements would lead to rapid evolution and diversification of MuseNet's capabilities.

Environmentally Conscious Training

Forecast: Given the significant energy requirements of training advanced models such as MuseNet, there might be a push toward more sustainable and efficient training methods.

Impact: Eco-friendly AI research would reduce the carbon footprint of models such as MuseNet, making them more sustainable in the long run.

Addressing Biases and Cultural Sensitivities

Forecast: Given the importance of cultural nuances in music, efforts would be made to address any biases in MuseNet's outputs and ensure it respects and understands cultural significance.

Impact: This would lead to more inclusive and sensitive AI-generated music, fostering global appreciation and understanding.

In conclusion, MuseNet's journey has only just begun. As the intersection of AI and music continues to evolve, MuseNet stands at the forefront of this exciting confluence. The future promises enhancements in technical capabilities, a deeper integration into our musical lives, and new avenues that we might not have even envisaged yet. However, with great potential comes the responsibility to navigate the ethical and societal implications carefully, ensuring that the harmony between man, machine, and music remains melodious.

18

Generating Images with BigGAN

Table of Contents

Getting to Know BigGAN

BigGAN, a name derived from its expansive architecture, represents one of the most significant advancements in the space of generative adversarial networks (GANs). Developed by DeepMind, its primary distinction lies in its ability to generate high-resolution and intricately detailed images, a leap that has often been described as remarkable in the annals of AI. Generative adversarial networks operate on a simple principle of having two neural networks, a generator and a discriminator, working in tandem. The generator creates images, while the discriminator evaluates them, acting as a critic. Their dynamic is akin to a forger trying to create a painting and a detective trying to spot its inauthenticity. Over time, the forger (generator) becomes so skilled that the detective (discriminator) struggles to distinguish the forgery from the original.

The genius of BigGAN is in its scale. Traditional GANs, when tasked with generating high-resolution images, often faced hurdles. The outputs might have been blurry, lacked coherence, or simply weren't detailed enough. BigGAN, by scaling up everything—the depth, width, and batch size of the network—managed to overcome many of these challenges. This *bigness* wasn't just about size but also about computational power and capacity. Training a model such as BigGAN requires considerable computational resources, something that's been a point of contention among critics concerned about the environmental impact of such large-scale models.

One of the intriguing aspects of BigGAN is its class-conditional generation. Instead of generating random images, BigGAN takes a more directed approach. It can be provided with a class label alongside its noise vector, instructing it on what kind of image to produce. This means if a researcher or artist wanted a specific category of image from a dataset, say an eagle from ImageNet, BigGAN would generate a high-resolution image of an eagle. This capability sets it apart from many of its contemporaries.

Yet, while its prowess is evident in the images it produces, BigGAN is not without its flaws. Sometimes, the images, although high in resolution, can possess surreal or unnatural elements. A bird might have too many wings, or a landscape might blend desert and icy terrains in ways not seen in nature. These quirks, though artistically intriguing, underscore the complexities and unpredictabilities inherent in training GANs. They serve as reminders that while the model might be learning from vast datasets, it doesn't truly *understand* the content in the way humans do. It's synthesizing based on patterns, not real-world knowledge.

Furthermore, the rise of tools such as BigGAN has sparked ethical debates in the AI community. The power to generate high-quality images opens the door to misuse, from creating misleading imagery to crafting deepfakes. As with all powerful tools, the benefits come with potential pitfalls.

In the grand tapestry of AI's evolution, BigGAN is a testament to how far GANs have come. It signifies a future where the synthesis of realistic media could be done at the click of a button. For artists, designers, filmmakers, and researchers, BigGAN offers a glimpse of a future where their creative and investigative horizons could be significantly expanded. However, like all powerful technologies, it comes with the imperative of responsible use, ethical considerations, and an awareness of its broader impact on society.

How Prompts Influence BigGAN

Guided Image Generation

At its core, BigGAN is designed for class-conditional generation. This means that instead of spewing out random images, BigGAN generates visuals based on specific class labels it's provided with. Prompts, in this context, function as these labels. For example, prompting BigGAN with "sunset over a mountain" can steer the neural network to produce an image reflecting this theme.

Refinement of Details

The specificity of a prompt can directly influence the granularity and nuance of the generated image. A vague prompt such as "bird" might lead to a generic image of a bird. In contrast, a more detailed prompt such as "a crimson-colored bird with elongated tail feathers" nudges BigGAN toward generating a more detailed and specific bird image, demonstrating the model's prowess in handling intricate instructions.

Facilitating Creative Exploration

While BigGAN is primarily a tool geared for realistic image generation, the prompts can also be used to explore abstract or novel concepts. Feeding it unconventional or surreal prompts can lead to the creation of unique, previously unseen images. For example, "a two-headed giraffe with butterfly wings" could generate a fantastical image, showcasing the model's capacity for creative exploration beyond real-world representations.

Batch Generation and Variability

BigGAN's architecture allows for the generation of multiple images in a single batch. By using prompts in tandem with varied noise vectors, users can obtain multiple interpretations or variations of a single idea, offering a spectrum of visual outputs for a singular concept.

Challenging the Model's Boundaries

Pushing BigGAN's capabilities through prompts can sometimes result in unexpected outcomes. These can range from stunningly detailed images to bizarre, unrealistic combinations. For example, prompting it with paradoxical or contradictory terms ("a freezing hot desert" or "a dry ocean") tests the model's limits and can produce intriguing results.

Overcoming Biases and Dataset Limitations

Every machine learning model carries with it biases from its training data. By carefully crafting prompts, users can attempt to steer BigGAN away from potential biases or stereotypical representations present in its training dataset, leading to more diverse and inclusive image outputs.

Enhancing Iterative Design

For designers or artists, prompts offer a way to iteratively refine their concepts. By gradually tweaking the prompt's wording or adding details, they can guide BigGAN toward their envisioned output, using the model as a collaborative tool in the design process.

Addressing Real-World Applications

In commercial or real-world scenarios, prompts can be employed to generate specific content tailored to niche requirements. Whether it's for marketing materials, concept art for gaming, or fashion design drafts, detailed prompts ensure BigGAN's outputs align with specific project needs.

Introducing Constraints for Experimentation

Sometimes, creativity thrives under constraints. By imposing specific limitations or conditions via prompts ("an urban landscape devoid of people" or "a forest with geometric trees"), users can challenge BigGAN to come up with unique visual solutions within set parameters.

Feedback Loop and Model Retraining

The results obtained from specific prompts can provide invaluable feedback on BigGAN's performance. Analyzing where the model excels or falters based on prompt-driven outputs can guide future model refinements and retraining efforts.

In conclusion, prompts play a crucial role in harnessing the full potential of BigGAN. They act as guiding beacons, influencing the direction, specificity, and creativity of the generated images. Whether it's to obtain a precise visual representation, explore artistic realms, or challenge the model's capabilities, prompts remain an essential tool in shaping BigGAN's outputs. As we continue to refine and understand the intricacies of generative models, the symbiotic relationship between prompts and the neural network will only grow in importance.

Practical Examples of BigGAN

Art and Digital Design

Description: Artists and designers can leverage BigGAN to create stunning visuals for their projects. Whether it's for album covers, poster designs, or digital art exhibitions, BigGAN can generate images tailored to specific themes or aesthetics.

Example: An electronic music artist might use BigGAN to produce abstract, cyberpunk-themed visuals for an upcoming album, ensuring a unique, AI-generated cover that resonates with the music's vibe.

Gaming and Entertainment

Description: The video game industry can use BigGAN to conceptualize environments, characters, or assets. It's a way to visualize ideas before they're developed, speeding up the preproduction phase.

Example: Game developers creating an alien-themed game can prompt BigGAN with descriptions of extraterrestrial landscapes or creatures, getting inspiration for levels or character designs.

Fashion and Apparel

Description: Fashion designers can use BigGAN for conceptualizing new patterns, textures, or designs for clothing. It can serve as a tool for brainstorming and visual inspiration.

Example: A fashion brand looking to launch a new summer collection might use BigGAN to generate tropical or beach-themed patterns for their fabrics, ensuring fresh and unique prints.

Research and Education

Description: Researchers can use BigGAN for visual simulations in fields such as biology, astronomy, or geology. Educators can employ it to generate visual aids for more engaging teaching.

Example: A biology teacher could use BigGAN to generate images of extinct creatures or hypothetical evolutionary paths, making classroom discussions more interactive and stimulating.

Advertising and Marketing

Description: Ad agencies can leverage BigGAN to craft compelling visuals for campaigns. It allows brands to stand out with unique, AI-generated content that captures attention.

Example: A beverage company launching a new tropical drink might use BigGAN to produce lush, vibrant jungle scenes as backdrops for their ads.

Architecture and Urban Planning

Description: Architects and urban planners can use BigGAN to visualize potential landscapes or cityscapes, aiding in the design and decision-making processes.

Example: Urban planners envisioning a sustainable city of the future might prompt BigGAN with terms such as "green rooftops," "solar-paneled roads," or "vertical gardens" to get potential visualizations.

Film and Animation

Description: Directors and animators can use BigGAN for concept art, storyboarding, or even creating backgrounds and assets for animated features.

Example: For a sci-fi film, BigGAN could generate dystopian or futuristic cityscapes, providing visual references for set designers and CGI teams.

Web and User Interface Design

Description: Web designers can use BigGAN to craft unique visuals, backgrounds, or icons for websites and applications, ensuring a distinct online presence.

Example: A travel website might use BigGAN to create dreamy, AI-generated landscapes, enticing users to explore more.

Medical Imaging and Simulations

Description: While BigGAN isn't a medical tool per se, its capacity to generate detailed images can aid in training and simulations, providing visual aids for medical professionals.

Example: Medical trainers could use AI-generated images to simulate rare conditions or scenarios, preparing students for a variety of cases.

Social Media and Content Creation

Description: Content creators can use BigGAN to craft unique visuals for their blogs, YouTube channels, or social media posts, offering fresh content to their followers.

Example: A travel blogger might use BigGAN to produce fantastical landscapes based on descriptions from ancient myths, blending history and imagination for their audience.

In essence, BigGAN serves as a bridge between human creativity and computational prowess. Its capacity to generate high-resolution, diverse images based on prompts makes it an invaluable asset across various industries, aiding in visualization, inspiration, and content creation. As AI continues to evolve, BigGAN's practical applications will undoubtedly expand, intertwining technology and artistry even further.

Challenges and Critiques of BigGAN

Computational Intensity

Description: One of the most pressing issues with BigGAN is its demand for computational power. Generating high-resolution images requires an immense amount of processing strength and memory.

Implication: This makes it less accessible for independent developers or small organizations. Only those with significant computational resources can harness the full potential of BigGAN without prohibitive delays or costs.

Training Data Concerns

Description: Like all machine learning models, BigGAN is only as good as the data it's trained on. There's always a risk that biases in the training data could be reflected in the outputs.

Implication: If the training data is not diverse enough, BigGAN might produce images that are stereotyped, biased, or not representative of real-world diversity.

Lack of Fine Control

Description: While BigGAN can generate images from prompts, the exact nature of the generated images can be unpredictable. Users don't have pixel-level control over the output.

Implication: This makes it challenging for tasks that require precision. For example, a designer might get an aesthetically pleasing image but not precisely what they had envisioned.

Ethical and Misuse Concerns

Description: The ability of BigGAN to create realistic images poses ethical challenges. There's potential for misuse in generating fake or misleading images.

Implication: In a world grappling with issues such as deepfakes and misinformation, the misuse of BigGAN in crafting deceptive visuals for malicious purposes is a genuine concern.

Overemphasis on Quantity over Quality

Description: While BigGAN excels at producing a vast array of diverse images, there's no guarantee on the quality or relevance of each image to the given prompt.

Implication: This could lead to scenarios where users have to sift through many generated images to find one that truly matches their needs.

Environmental Impact

Description: Training deep learning models such as BigGAN has a significant carbon footprint due to the vast computational resources required.

Implication: As AI research grows and models become more complex, the environmental impact of training these models becomes a concern for sustainable AI development.

Lack of Creativity and Originality

Description: While BigGAN can generate unique images, critics argue that its creations are remixes of its training data, lacking true originality.

Implication: This raises questions about the value of AI-generated art and its place in creative industries. Can BigGAN-produced images ever match human creativity?

Dependency and Over-Reliance

Description: As tools such as BigGAN become more integrated into industries, there's a risk of over-reliance, where human skills might get sidelined.

Implication: This could lead to reduced emphasis on human creativity, intuition, and design skills in favor of machine-generated content.

Lack of Interpretability

Description: Deep learning models, including BigGAN, are often termed *black boxes* because their internal workings are not entirely understood, even by their developers.

Implication: This lack of transparency can be concerning, especially when we don't fully understand why a model produces a particular output.

Economic and Job Implications

Description: With the increasing capabilities of models such as BigGAN, there's potential for automation in sectors such as design, leading to economic implications.

Implication: If machines can produce art, designs, or visuals at a fraction of the time and cost, it might affect job opportunities and economic structures in creative fields.

In conclusion, while BigGAN offers groundbreaking capabilities in image generation, it's not without its challenges and critiques. Balancing its potential with ethical, environmental, and economic considerations will be crucial as we move forward in the AI-driven future. As with all tools, the key lies in understanding its limitations and using it responsibly, complementing human creativity rather than replacing it.

The Future of BigGAN

Scalability and Efficiency

Expansion of computational capabilities: As we move into an era of advanced chip designs and better GPUs, running models such as BigGAN will become more feasible for a broader audience.

Model pruning and optimization: Future iterations of BigGAN may focus on refining the model's architecture to make it more lightweight without compromising on its image generation quality.

Enhanced Fine-Tuning and Precision

Prompt precision: Advances in model training might enable more precise image generation, granting users better control over specifics such as color, texture, and form.

Feedback loops: Implementing human-in-the-loop training systems could refine BigGAN's outputs based on user feedback, ensuring more aligned and desirable results.

Integration with Augmented Reality (AR) and Virtual Reality (VR)

Dynamic content creation: BigGAN could play a role in generating real-time, high-resolution visuals for VR and AR environments, offering immersive experiences.

Interactive design: As VR and AR design tools become more sophisticated, BigGAN might provide instant visual prototypes based on user instructions, facilitating creative processes.

More Diverse and Ethical Training Data

Anti-bias frameworks: Given the concerns about biases in AI, future BigGAN iterations may emphasize sourcing diverse and representative training datasets.

Transparency initiatives: Open sourcing training data or providing detailed data documentation can build trust and ensure ethical image generation.

Environmental Considerations

Green AI: In light of the environmental impact of training large models, there may be a push toward more sustainable practices, for instance, using renewable energy for data centers.

Optimized training: Techniques such as knowledge distillation or transfer learning can reduce the training time and resources required, diminishing the carbon footprint of models such as BigGAN.

Broader Accessibility and Democratization

Platform integration: We might see platforms or apps offering "BigGAN-as-a-service," enabling designers, artists, and developers to generate images without delving into the complexities of the model.

Educational initiatives: To increase AI literacy, there could be courses or workshops focusing on using BigGAN for various domains, from digital art to scientific visualization.

Enhanced Collaboration with Human Creativity

AI-assisted design: Rather than replacing human designers, BigGAN can act as a tool in the designer's arsenal, suggesting ideas or rapidly prototyping visuals.

Creative exploration: Artists might use BigGAN as a muse, allowing the model to generate a base image, which they can then modify or interpret, fostering a symbiotic relationship between man and machine.

Addressing Ethical and Misuse Concerns

Digital watermarking: To combat potential misinformation, images generated by BigGAN could be embedded with watermarks or metadata to denote their AI-generated nature.

Regulatory frameworks: As AI-generated content becomes more prevalent, there may be legal and regulatory frameworks developed to ensure responsible usage and to penalize malicious applications.

Commercial Applications and Monetization

Industry-specific solutions: Different sectors, from gaming to fashion, could have BigGAN versions tailored to their specific needs, generating relevant content.

Licensing models: Given the potential of BigGAN to create unique visuals, there might be platforms allowing artists to sell or license their AI-generated artwork.

Push toward Generality and Multimodality

Cross-media synthesis: The future might see a fusion of models such as BigGAN with text or audio generators, leading to a unified system that can produce multimedia content.

Understanding context: Enhanced training methods could lead BigGAN to not just generate images based on prompts but understand the context behind them, making image generation more holistic and meaningful.

In summary, the horizon for BigGAN is vast and laden with possibilities. While its current capabilities are already transformative, the confluence of technological advancements, ethical considerations, and human creativity promises an exciting future. The interplay of BigGAN with various domains, from art to industry, will shape its evolution, driving it toward a future that's both inclusive and innovative.

19

Creating Code with Codex

Table of Contents

Introduction to Codex

Codex represents not merely the next iteration of code-generation tools but an avant-garde transformation of how we perceive the confluence of coding and AI. At its essence, Codex is a prodigious language model specifically designed for generating code, and its inception can be considered both an evolution and a revolution.

Codex's lineage traces back to the landmark achievements of its predecessors, particularly GPT-3, one of Open AI's most lauded creations. While GPT-3 demonstrated awe-inspiring abilities across a wide range of natural language processing tasks, Codex fine-tunes this prowess with an emphasis on code generation. The very architecture of Codex is testament to OpenAI's commitment to pushing the boundaries of what AI models can achieve. Trained on a rich amalgamation of licensed data, data crafted by human trainers, and a plethora of publicly available data, Codex emerges as a testament to the sheer potential of deep learning in understanding and generating programming languages.

But Codex isn't just about transforming programming paradigms; it's also about accessibility. Recognizing the global developer community's vast and varied needs, OpenAI facilitated access to Codex via a dedicated API. This move was strategic and far-reaching, allowing developers, irrespective of their niche, to integrate Codex's capabilities seamlessly into their applications, tools, and services. The API wasn't just a gateway to Codex's power but a testament to OpenAI's vision of making avant-garde AI advancements accessible and utilitarian.

A key aspect that sets Codex apart is its polyglot nature. In the realm of programming, where numerous languages exist, each with its unique syntax, semantics, and applications, Codex's ability to traverse multiple languages is nothing short of revolutionary. From the versatile Python and the ubiquitous JavaScript to even more niche languages, Codex's spectrum of understanding and generation spans wide, affirming its role as a universal coding assistant.

However, beyond its technical specifications and capabilities, Codex stands as a symbol of the evolving relationship between human developers and AI. It prompts us to reimagine coding not as a solitary human endeavor but as a collaborative process where man and machine work in tandem. Codex's introduction marks a pivotal moment where we transition from viewing AI as a tool to seeing it as a partner. This partnership, laden with potential, could redefine software development, debugging, and even education.

Of course, as with any groundbreaking innovation, Codex invites its fair share of scrutiny and contemplation. The ethics of automated code generation, the implications for novice developers, and the broader consequences on the software development industry are debates that Codex invariably ignites. However, these conversations are not just challenges but opportunities—opportunities to mold Codex's trajectory in a manner that is both ethically sound and technologically progressive.

In summation, Codex is not just a code-generating AI; it's a beacon of what's possible at the intersection of deep learning and software development. As developers, educators, and tech enthusiasts familiarize themselves with Codex, they aren't just engaging with a tool; they're interacting with the future of coding. A future where the lines between human creativity and machine efficiency blur, where challenges are met with collaborative solutions, and where the very act of coding becomes a harmonious dance between human intent and AI capability. The introduction of Codex is more than a technological milestone; it's a clarion call for an era of collaborative coding.

The Impact of Prompts on Codex

Human-AI Interaction Interface Prompts serve as the primary means of communication between the user and Codex. Instead of interacting through complex configurations or abstract code, prompts allow developers to convey their requirements in natural language. This intuitive mode of interaction makes Codex accessible not only to seasoned developers but also to individuals with limited coding experience.

Contextual Understanding The power of Codex isn't just in code generation but in its ability to discern context from the given prompts.

For example, when a user specifies, "Create a Python function to filter even numbers from a list," Codex understands the language preference, the nature of the function, and the specific operation to be performed. This contextual comprehension ensures that the generated code aligns with the user's intention.

Adaptive Code Generation Depending on the specificity or vagueness of the prompt, Codex can generate a wide range of code outputs. A generic prompt might elicit a template-based response, while a more detailed prompt can yield a highly customized piece of code. This adaptability ensures that Codex can cater to a diverse array of coding needs and scenarios.

Error Identification and Rectification Codex's interaction with prompts isn't one-dimensional. When given a piece of code and prompted to identify errors, Codex can pinpoint issues and suggest corrections. This feedback loop, initiated by prompts, transforms Codex from a passive code generator to an active debugging assistant.

Facilitating Learning and Education For learners and educators, prompts serve as a bridge to understanding complex coding concepts. By prompting Codex with queries such as "Explain the concept of recursion in coding" or "Demonstrate a binary search algorithm," users can receive not only code snippets but also explanations and annotations. This educational potential democratizes coding knowledge, making it accessible to a broader audience.

Enhancing Productivity in Development Codex, when prompted aptly, can generate boilerplate code, set up standard structures, or even produce intricate algorithms, thereby accelerating the software development process. Instead of starting from scratch, developers can use prompts to get a head start, refining the generated code as needed.

Multilingual Code Generation The versatility of Codex extends to its proficiency in multiple programming languages. Through prompts, users can specify their language of choice, be it Python, JavaScript, Java, or any other language Codex supports. This multilingual capability, driven by prompts, empowers developers to seamlessly switch between projects with varied language requirements.

Interactive Code Refinement Codex's interaction with prompts is iterative. If the generated code doesn't meet the user's expectations, they can refine their prompt or provide additional context to guide Codex toward the desired output. This dynamic interaction ensures that the final code aligns with the user's objectives.

Integration with Tools and Platforms The utility of Codex extends beyond stand-alone code generation. Through API-based prompts, Codex can be integrated into development environments, platforms, and tools. This seamless integration, facilitated by prompts, means that developers can harness Codex's capabilities directly within their preferred coding ecosystems.

Ethical and Responsible Coding As AI-generated code becomes more prevalent, there's a growing emphasis on ethical and responsible coding practices. Through carefully crafted prompts, developers can guide Codex to generate code that adheres to best practices, industry standards, and ethical guidelines.

In conclusion, prompts are more than mere instructions in the realm of Codex; they are catalysts that unleash its full potential. They foster a dynamic and symbiotic relationship between the user and the AI, ensuring that the generated code is not just technically sound but also contextually relevant. As Codex continues to evolve, the role of prompts will undoubtedly remain pivotal, steering the AI toward greater accuracy, versatility, and utility.

Real-World Use Cases and Examples

Rapid Prototyping

Explanation: Codex can be used to quickly generate prototypes or MVPs for applications. By providing high-level requirements through prompts, developers can get a foundational codebase in minutes.

Example: A start-up wishes to create a basic web app for user registration. A prompt such as "Generate a web app with a user registration and login system using Flask and SQLite" could result in Codex producing the necessary back-end code.

Educational Tool

Explanation: Codex serves as a valuable resource for students and educators, helping them understand coding concepts, algorithms, or data structures.

Example: A student struggling with the concept of a linked list could prompt Codex with "Explain and demonstrate a singly linked list in Python" to receive both a code sample and an explanatory note.

Integration with Development Platforms

Explanation: Codex can be integrated directly into IDEs or development platforms, assisting developers in real time as they work on their projects.

Example: Within a popular IDE, if a developer is unsure how to implement a specific function, they might type "Codex Create a function to merge two sorted lists" and receive an immediate response.

Code Refactoring

Explanation: Codex can help optimize and refactor existing code, making it more efficient or readable.

Example: Given a block of complex and tangled code, a developer could prompt Codex with "Simplify and refactor this Python code for clarity" to get a more streamlined version.

Bug Identification and Debugging

Explanation: Codex can be prompted to identify errors or potential bugs in a piece of code and suggest fixes.

Example: Upon encountering an error in a script, a developer might input the problematic code section and prompt Codex with "Identify and fix errors in this Python script."

Code Translation between Languages

Explanation: Codex can assist in translating code snippets from one programming language to another, saving developers time and ensuring accuracy.

Example: A developer has a JavaScript function that they wish to reproduce in Python. They could prompt Codex with "Translate this JavaScript function to Python" to receive an equivalent Python function.

Guided Code Development with Natural Language

Explanation: Developers can instruct Codex in plain English (or other languages) to generate code without needing to specify every detail.

Example: An app developer wishes to add a feature but isn't sure about the syntax. They might prompt, "Add a feature to this mobile app that notifies users every morning at 8 a.m."

Assisting in Database Operations

Explanation: Codex can generate SQL queries or scripts based on descriptive prompts, assisting in database operations.

Example: A database administrator wants to extract specific data but isn't sure of the optimal query. They could prompt Codex with "Generate an SQL query to find all users from the 'users' table who joined after January 1, 2020."

Custom Code Snippets and Libraries

Explanation: Codex can assist in creating custom functions, classes, or libraries tailored to specific requirements.

Example: A game developer wishes to create a specific movement mechanic. They might prompt Codex with "Write a function in C# for a double jump mechanic in Unity."

Documentation and Code Comments

Explanation: Codex can be used to generate explanatory comments or documentation for existing code, enhancing readability and maintainability.

Example: Given a complex algorithm, a developer could prompt Codex with "Provide a detailed comment explaining this sorting algorithm."

In the evolving landscape of software development, Codex emerges as a versatile ally. From rapid prototyping to real-time debugging, its potential applications span the entire development life cycle. By understanding and leveraging these real-world use cases, developers and organizations can harness the power of Codex to streamline processes, enhance code quality, and catalyze innovation.

Limitations and Areas for Improvement

Complexity Limit

Explanation: While Codex is robust, it may sometimes struggle with understanding or generating intricate systems or code requirements. This limitation is especially evident when dealing with highly specialized or niche development tasks.

Dependence on Clear Prompts

Explanation: The efficacy of Codex largely hinges on the clarity and specificity of user prompts. Vague prompts can lead to inaccurate or suboptimal code generation. Users need to iterate and refine their questions to get desired outputs.

Potential for Code Inefficiency

Explanation: Codex might not always produce the most optimized or efficient code. While it might correctly perform a requested task, the code might not be as refined or performance-tuned as a solution handcrafted by an experienced developer.

Risk of Reinforcing Bad Practices

Explanation: If Codex is trained on datasets that include poorly written or unoptimized code, there's a risk it might inadvertently promote or reinforce bad coding practices.

Inadequate Handling of Edge Cases

Explanation: Codex might not always account for all potential edge cases in the solutions it provides. This could lead to software that works under most conditions but fails under specific circumstances.

Security Concerns

Explanation: Automatically generated code could potentially introduce vulnerabilities if not reviewed properly. Additionally, sharing proprietary or sensitive code snippets with cloud-based instances of Codex could raise data privacy and intellectual property concerns.

Over-Reliance and Skill Degradation

Explanation: An over-reliance on tools such as Codex might lead to a potential degradation in coding skills among developers. If they begin to rely solely on Codex for solutions, they may not practice or develop their problem-solving abilities as much as they should.

Lack of Domain-Specific Knowledge

Explanation: Codex operates based on its training data. When asked about domains or technologies that are too recent, niche, or specialized, it might not have sufficient knowledge or context to generate accurate code.

Integration Challenges

Explanation: Integration of Codex into existing development environments and workflows might present challenges. It may not always be straightforward, especially in proprietary or highly customized development settings.

Ethical and Job Market Implications

Explanation: The automation of code creation and debugging tasks might raise concerns about job displacement in the software development industry. While Codex is a tool meant to assist developers, there are debates about the long-term implications of such powerful automation tools on the job market.

Licensing and Copyright Issues

Explanation: As Codex generates code, questions arise about the ownership and licensing of that code. Organizations need to be aware of potential legal gray areas, especially when using generated code in commercial applications.

Cost and Accessibility

Explanation: Depending on its deployment, accessing or using Codex might come at a cost, which could hinder its widespread adoption, especially among individual developers or small start-ups.

Quality Assurance and Testing

Explanation: Generated code will still require rigorous testing. Relying on Codex might give a false sense of security, leading to potential oversights during quality assurance processes.

Potential for Misuse

Explanation: Like any powerful tool, Codex has the potential for misuse. Bad actors might utilize Codex to rapidly generate malicious code or scripts.

Overwhelming Choices

Explanation: Given its vast knowledge base, Codex might occasionally provide multiple approaches to a single problem. This could potentially overwhelm or confuse developers who are seeking a single, definitive solution.

Codex represents a significant leap forward in the domain of code generation and assistance. However, understanding its limitations is crucial for developers and organizations aiming to incorporate it into their workflows. Proper training, rigorous review processes, and a balanced reliance will ensure that Codex serves as an effective tool, augmenting human capabilities rather than replacing or undermining them.

Looking Forward: The Future of Codex

Integration with Development Environments

Insight: Codex's capabilities will likely become deeply incorporated into popular integrated development environments (IDEs). This seamless incorporation will allow developers to get real-time suggestions, error fixes, and alternate coding approaches as they type, enhancing the development process.

Custom Training and Specialization **Insight:** Future versions of Codex might offer users the ability to fine-tune the model on their codebases. This means Codex could learn an organization's coding style, preferred practices, and even domain-specific intricacies, making it an invaluable personalized coding assistant.

Enhanced Multimodal Capabilities **Insight:** With the growth of multimodal models, Codex could evolve to understand and generate code not just from textual prompts but also from diagrams, voice commands, or other non-textual inputs, making it even more versatile.

Improved Efficiency and Optimization

Insight: As feedback on Codex's generated code accumulates, we can expect the model to produce increasingly optimized and efficient code. The AI will learn from developer modifications to its suggestions, leading to better and more refined code outputs over time.

Expanded Language and Framework Support **Insight:** While Codex already supports a wide range of programming languages and frameworks, it's likely that this support will expand. It could encompass more niche languages, newer frameworks, and even upcoming programming paradigms.

Collaborative Code Generation **Insight:** Codex might evolve to support real-time collaborative coding, where multiple developers and the AI work in tandem. It could act as a mediator, ensuring consistency in style, spotting potential integration issues, and providing solutions during collaborative sessions.

Enhanced Security and Privacy Measures **Insight:** Given concerns about sharing proprietary code or potential security vulnerabilities in generated code, there will likely be a strong emphasis on enhancing Codex's security. This includes better data handling practices and possibly the ability to run Codex entirely offline in a secure environment.

Education and Training **Insight:** Codex can be a powerful tool for education. Future iterations could be used to assist students in learning programming, offering real-time feedback, hints, and explanations, making the learning curve less daunting for newcomers.

Broader Industry Applications **Insight:** Beyond software development, Codex's capabilities could be extended to other industries such as bioinformatics, engineering, or finance where scripting and coding play crucial roles. Codex could assist professionals in these sectors by automating routine tasks or offering solutions to domain-specific problems.

Natural Language Enhancement **Insight:** As Codex is a language model at its core, its understanding of natural language will continue to improve. This will result in better interpretations of user prompts, even if they are vague or ambiguous, leading to more accurate code generation.

Ethical and Regulatory Developments **Insight:** As with all powerful AI tools, there will be discussions and potential regulations concerning Codex's impact on the job market, intellectual property rights, and software quality. This could shape how Codex is used, especially in critical applications.

Code Review and Testing Assistance **Insight:** Beyond generating code, Codex could play a role in the code review and testing phases. It might offer insights into potential vulnerabilities, suggest test cases based on the generated code, or even predict how changes might affect existing systems.

Sustainable AI Development **Insight:** Given the environmental concerns associated with training large models, future iterations of Codex might focus on more energy-efficient training methods, smaller model sizes, or techniques that require less computational power.

Increased Accessibility

Insight: To make Codex more accessible to individual developers, hobbyists, or smaller start-ups, there might be efforts to offer more cost-effective or even open-source versions, democratizing access to advanced AI-assisted coding.

Global and Cultural Adaptation **Insight:** Codex might evolve to cater to global audiences better, understanding coding nuances, comments, or prompts in multiple languages and cultural contexts, ensuring a more inclusive development environment.

In summary, Codex's journey, although impressive so far, is just beginning. Its potential to reshape the software development landscape is immense, but its success will depend on how well it adapts to user needs, industry changes, and the ever-evolving tech ecosystem.

20

Generating 3D Art with RunwayML

Table of Contents

Introduction to RunwayML

RunwayML, a groundbreaking platform in the contemporary digital art space, has emerged as a game changer for artists, designers, and innovators alike. It stands as a testament to the intersection of art and technology, particularly in how machine learning can redefine the boundaries of creative expression. This introduction delves into the origin, capabilities, and transformative potential of RunwayML in the vast expanse of digital artistry.

Founded with the vision of democratizing AI's artistic potential, RunwayML has positioned itself as an indispensable tool for those wishing to venture into the domain of 3D art without being bogged down by the complexities of traditional machine learning pipelines. The platform, rather than being a mere tool, can be viewed as a bridge—a bridge that connects the vast and often intimidating world of deep learning with the more intuitive and subjective realm of artistic creativity.

At the heart of RunwayML is its ability to utilize pretrained models, particularly generative adversarial networks (GANs), which are renowned for their aptitude in generating realistic content. These models, traditionally requiring robust computational power and expertise to operate, are made accessible via a user-friendly interface within RunwayML. This means that even those without a deep technical background can harness the power of advanced algorithms to generate intricate 3D artworks.

However, RunwayML isn't solely about simplifying the technical. Its true genius lies in how it reimagines the role of machine learning in the creative process. Instead of positioning technology as a mere tool, RunwayML invites it to be a collaborator. Through the platform, artists provide prompts or set parameters, and the algorithm responds, leading to a dynamic interplay between human intuition and machine-generated suggestions. This synergy results in art that is neither purely human nor purely machine but a fusion of both, pushing the boundaries of what is conventionally deemed possible.

The adaptability of RunwayML is also worth noting. It isn't pigeonholed into a specific genre of art or design. From creating abstract visualizations and surreal landscapes to sculpting lifelike 3D models, the platform's versatility caters to a broad spectrum of artistic endeavors. Furthermore, its potential isn't just limited to static artworks. The rise of augmented and virtual reality has opened up a plethora of opportunities for dynamic 3D content, and RunwayML stands poised to play a pivotal role in shaping these immersive digital experiences.

While its capabilities are undeniably impressive, the true essence of RunwayML is rooted in its philosophy. In an era where the debate around machine learning often gravitates toward automation and job displacement, RunwayML offers a refreshing narrative. It emphasizes collaboration over replacement and augments human creativity rather than diminishing it. This ethos, combined with its technical prowess, positions RunwayML not just as a software platform but as a movement—one that champions the harmonious coexistence of art and technology.

In sum, RunwayML, with its fusion of cutting-edge machine learning and user-centric design, emerges as more than a mere digital tool. It represents a paradigm shift in how we perceive the relationship between artists and technology, serving as a beacon for a future where machines don't just aid the creative process but actively participate in it. As artists and designers continue to explore and push its boundaries, RunwayML promises to reshape the canvas of digital artistry in unprecedented ways.

The Role of Prompts in RunwayML

Initiating Artistic Direction

Definition: Prompts act as initial seeds or guidelines provided by the user to steer the algorithm toward a specific artistic direction.

Explanation: In the vast potential landscape of machine-generated art, prompts serve as the user's way of guiding the system. It's akin to giving a brief to a fellow artist. By providing a prompt, creators can ensure that the final output aligns more closely with their envisioned idea.

Interactivity and Real-Time Feedback

Definition: RunwayML facilitates real-time interactions, enabling users to modify prompts and witness instantaneous changes in the output.

Explanation: This dynamic process of interaction empowers artists. They can experiment with different prompt combinations, tweak parameters, and immediately see how these changes influence the final art piece. This live feedback loop fosters experimentation and helps in refining the artwork to its finest detail.

Contextual Understanding

Definition: Prompts provide the necessary context to RunwayML, enabling it to understand the desired mood, style, or theme for the artwork.

Explanation: Machine learning models, by default, lack an intrinsic understanding of art. However, when fed with the right prompts, they can generate outputs that capture the essence of specified themes, be it surrealism, Renaissance, or even specific inspirations such as "a starry night over a serene lake."

Collaborative Art Creation

Definition: Prompts establish a symbiotic relationship between the artist and the machine, leading to collaborative art generation.

Explanation: Instead of the artist being a passive observer, the use of prompts in RunwayML makes them an active participant. They're not just consuming machine-generated content but are co-creating with the system. This partnership allows for outputs that are unique amalgamations of human creativity and algorithmic possibilities.

Bridging Skill Gaps

Definition: Prompts act as a bridge for those who might not possess advanced technical knowledge in 3D art or machine learning.

Explanation: With the help of intuitive prompts, individuals who might not have deep expertise in 3D modeling or design can still produce high-quality artworks. This democratizes the art creation process, making it accessible to a wider audience.

Enhancing Iterative Design

Definition: Prompts support the iterative design process, allowing creators to revisit and modify their inputs based on previous outcomes.

Explanation: Art and design are often iterative. An artist might not get the desired result in the first attempt. With RunwayML, after observing an initial output, artists can adjust their prompts to guide the algorithm closer to their envisioned result in subsequent iterations.

Diverse Artistic Outputs from Single Inputs

Definition: A single prompt can yield a multitude of diverse artistic outputs based on subtle changes or interpretations.

Explanation: Given the inherent randomness and vast potential of generative algorithms, even a slight tweak in a prompt or its parameters can lead to drastically different artistic creations. This variability is a boon for artists seeking diverse interpretations of a singular idea.

In conclusion, prompts in RunwayML aren't merely textual inputs but are foundational to the entire creative process on the platform. They serve multiple roles, from guiding the algorithm and fostering interactivity to facilitating collaborative art creation. In the ever-evolving realm of machine-assisted art, prompts act as the vital communication link, ensuring that the human artist's vision is not just preserved but amplified in the final masterpiece.

Showcasing Real-World Examples

Interactive Art Installations

Description: Art galleries and exhibitions have started to use RunwayML to create interactive art installations that respond to visitors' movements or gestures.

Explanation: For instance, a particular installation might consist of a projected 3D landscape that changes its topology and color schemes based on the audience's proximity or hand gestures. This not only makes art more engaging but also adds an extra dimension of personalization for every visitor.

Fashion and Apparel Design

Description: Designers are harnessing RunwayML to visualize futuristic fashion concepts in 3D before they are physically created.

Explanation: By inputting certain keywords or style inspirations into the system, designers can see a 3D representation of a clothing item. This aids in faster prototyping, reducing material waste, and exploring avant-garde designs that might be too risky to create without a prior visual.

Architectural Visualizations

Description: Architects and urban planners use RunwayML to generate 3D visualizations of buildings, landscapes, or entire city blocks.

Explanation: For example, if a planner wants to envision a "sustainable city block with green rooftops and pedestrian zones," inputting this prompt can generate a 3D model. This accelerates the design process and helps stakeholders visualize end results more effectively.

Gaming and Virtual Worlds

Description: Game developers employ RunwayML to design 3D characters, environments, and objects for video games.

Explanation: Instead of manually modeling each element, developers can provide prompts such as "a post-apocalyptic city" or "a mystical forest creature" to get varied 3D models. These can then be fine-tuned and integrated into games, enhancing the richness and diversity of virtual worlds.

Film and Animation

Description: Filmmakers and animators utilize RunwayML for conceptualizing scenes, characters, or entire sequences in their projects.

Explanation: Consider a filmmaker wanting to visualize a "futuristic city at dawn with flying cars." Instead of drawing or manually modeling this scene, they can use RunwayML to generate a 3D draft, aiding in set design, storyboarding, and production planning.

Educational Tools

Description: Educators use RunwayML to create 3D representations of abstract concepts or historical events.

Explanation: For example, a history teacher might generate a 3D model of "Ancient Rome's marketplace" to give students a more immersive understanding of historical contexts. This visual approach can enhance comprehension and retention for learners.

Advertising and Marketing Campaigns

Description: Brands are turning to RunwayML to develop unique 3D visuals for their advertising campaigns.

Explanation: A sports brand, for instance, might generate 3D visuals of "futuristic athletic gear in a neon-lit urban setting" for a campaign targeting younger audiences. These captivating visuals can make campaigns more memorable and powerful.

Product Prototyping

Description: Companies use RunwayML to visualize potential product designs in 3D before physical manufacturing.

Explanation: A furniture company could visualize a "minimalist wooden coffee table with asymmetrical legs" to gauge aesthetic appeal and practicality. This can inform decisions related to design modifications, material selection, and market viability.

Customizable Digital Art Platforms

Description: Several online art platforms now allow users to generate personalized 3D artworks using RunwayML.

Explanation: Users can provide prompts like "a tranquil mountainscape with autumn colors" and receive a unique 3D artwork. This democratizes art creation and allows non-artists to express their visions.

Museum and Heritage Sites

Description: Museums utilize RunwayML to recreate historical artifacts, settings, or events in 3D for virtual tours or augmented reality experiences.

Explanation: For example, a museum showcasing ancient civilizations might recreate "a bustling marketplace in Mesopotamia" in 3D. This enhances visitors' experiences, making history more tangible and engaging.

In essence, RunwayML's capabilities have permeated various sectors, from art and design to education and advertising. Its ability to rapidly generate detailed and varied 3D models based on simple textual prompts is revolutionizing how professionals and enthusiasts alike approach 3D design and visualization.

Strengths and Weaknesses

RunwayML, an avant-garde platform tailored for the intersection of machine learning and creative endeavors, has gained traction as a bridge connecting artists and AI. Leveraging a user-friendly interface, it facilitates the deployment of machine learning models in artistic projects, especially in the realm of 3D art generation. However, like any tool, it has its own strengths and weaknesses that can influence its adoption and usage in varied scenarios.

Strengths

User-Friendly Interface One of the most significant advantages of RunwayML is its intuitive, user-centric design. For artists and creators who may not have a technical background in machine learning or coding, this platform offers a seamless experience. By simplifying the complex operations, it has democratized access to cutting-edge AI technology.

Diverse Model Library The platform hosts an array of pretrained models, each catering to different artistic requirements. This allows creators to explore a vast expanse of possibilities, from style transfers to 3D model generation, without the need to train or understand the intricacies of each model.

Real-Time Interactivity RunwayML enables artists to interact with models in real time. This facilitates immediate feedback, allowing for iterative and dynamic creation processes. Artists can tweak parameters on the go and see the results instantaneously.

Cloud Integration The infrastructure of RunwayML is designed to leverage the cloud. This means that artists and creators don't necessarily need high-end computing hardware on their end. The heavy computational tasks are offloaded to the cloud, ensuring smooth performance regardless of the user's system specifications.

Collaboration and Community With features designed for collaboration, artists can work together on projects, share insights, and even integrate third-party applications. Moreover, the RunwayML community is burgeoning, fostering knowledge exchange and collaborative learning.

Weaknesses

Dependency on Internet Given its cloud-centric design, RunwayML requires a stable Internet connection. This could be a constraint for artists working in areas with erratic Internet or those who prefer offline tools for their creative processes.

Cost Implications While the platform offers a plethora of functionalities, some features and higher computational capacities come at a cost. For artists working on expansive projects or those who are frequent users, this could lead to significant expenses.

Limited Customizability Despite its extensive model library, there might be occasions when artists seek a specific output that isn't catered to by the preexisting models. While RunwayML is designed to accommodate a wide variety of artistic needs, it doesn't offer the depth of customization that might be achieved through bespoke model training.

Quality Variance The results produced by AI models can sometimes be unpredictable. Even with the same parameters, slight changes in input can lead to vastly different outputs. For artists seeking consistent results, this can be a challenge.

Learning Curve While designed to be user-friendly, any new platform comes with its learning curve. Artists new to the domain of AI or those transitioning from other platforms might require some time to acclimate to RunwayML's environment.

In summation, RunwayML emerges as a potent tool for artists looking to intertwine their creativity with the capabilities of AI. Its strengths make it an accessible and robust platform for a wide range of artistic endeavors. However, potential users should be aware of its limitations. By understanding these strengths and weaknesses, artists can make an informed decision on how to best integrate RunwayML into their creative workflows.

Future Directions for RunwayML

As we stand on the cusp of a new era in the marriage of creativity and machine learning, platforms such as RunwayML are poised to redefine the landscape of 3D art generation and beyond. The integration of AI with creative processes has just begun, and the trajectory of RunwayML, given its current capabilities and potential, hints at several future directions.

Expansion of Model Library One of the inherent strengths of RunwayML is its vast repository of pretrained models. As the field of machine learning witnesses rapid advancements, it's anticipated that RunwayML will incorporate newer models with advanced capabilities. This would mean a broader spectrum of artistic functionalities, ranging from more realistic 3D models to potentially integrating 4D elements, i.e. time, to create dynamic art pieces.

Integration with Augmented and Virtual Reality The world is gradually transitioning toward more immersive experiences. With the growth of AR and VR technologies, it's plausible for RunwayML to delve into these realms, offering artists tools to create 3D art optimized for AR and VR environments. This would amplify the experiential component of art, transforming passive viewers into active participants.

Enhanced Collaborative Features As the global community becomes more interconnected, collaborative art forms are gaining traction. Future versions of RunwayML might incorporate enhanced collaborative tools, allowing multiple artists from diverse geographical locales to work on a single project simultaneously. Imagine an art piece evolving in real time as artists from different continents contribute to it, all facilitated by RunwayML.

Custom Model Training While the current strength lies in its pretrained models, there's a growing demand for customizability in artistic expressions. RunwayML could venture into allowing artists to train their models, tailored to specific needs. This would cater to advanced users seeking a unique touch, unmatched by preexisting models.

Sustainability and Eco-Friendly Computation The computational demands of advanced machine learning models are high, leading to concerns about environmental impacts, especially in terms of carbon footprints. As part of its future trajectory, RunwayML could incorporate eco-friendly computational methods, such as energy-efficient algorithms or leveraging renewable energy for cloud computations. This would resonate with a growing segment of eco-conscious artists and consumers.

Enhanced Offline Capabilities Recognizing the challenges associated with consistent Internet dependencies, it's foreseeable for RunwayML to enhance its offline functionalities. This would ensure that artists in remote locations or those in areas with unstable Internet can still harness the platform's capabilities without hindrance.

Education and Training Modules With an expanding user base, there's an opportunity for RunwayML to integrate educational modules. These could range from basic tutorials for beginners to advanced masterclasses for seasoned artists. By fostering a culture of continuous learning, the platform can ensure that its users are always abreast of the latest features and best practices.

Open-Source Ecosystem Development Building a community-driven open-source ecosystem can propel RunwayML to new heights. By allowing developers and artists to contribute to the platform's growth, it can tap into the collective intelligence of a global community, leading to innovations that a closed system might not conceive of.

Personalized AI Assistants As AI becomes more sophisticated, we can envisage a future where RunwayML offers personalized AI assistants to artists. These assistants, learning from the individual artistic preferences and styles of users, could provide real-time suggestions, optimizations, and even anticipate artists' needs.

Integration with Other Art Forms While 3D art generation is a significant component, the future might see RunwayML branching out to integrate other art forms, such as music, literature, or dance. This multidisciplinary approach can lead to the birth of entirely new genres of art, driven by AI's capabilities.

In closing, RunwayML, at the intersection of creativity and technology, is primed for a future that promises uncharted artistic territories. The blend of AI's computational prowess with human creativity is still in its nascent stages, and platforms such as RunwayML are the torchbearers leading the way. As the canvas of possibilities expands, the future for RunwayML, and by extension, the world of 3D art, seems both exciting and limitless.

Sample Prompts for RunwayML

The general structure of a prompt such as the following is a great start.

```
[subject] in the style of [style or aesthetic].
```

Prompt Modes (Source: RunwayML)

1. **Text only:** Best for creating from only words. Gen-2 will generate footage matching your prompt.
2. **Image only:** Best for perfectly matching an image. Gen-2 will animate the image into a video. No image previews are provided since the image preview would just be your input image directly.
3. **Image and text description:** The video will match the image directly and use the text to help create the video.

For more information, check out Gen-2 Prompting Modes.

Text Prompt Tips The following generic prompt modifiers generally produce good results:

```
masterpiece
classic
cinematic
```

For more action and movement, try adding keywords such as:

```
cinematic action
flying
speeding
running
```

Try using camera-specific terms:
camera angles (`full shot`, `close up`, etc.)
lens types (`macro lens`, `wide angle`, etc.)
camera movements (`slow pan`, `zoom`, etc.)

In general, trying to explain camera movements as if you were explaining to a toddler can help; so instead of `dolly shot`, you would say something such as `zooming in over time` or `growing over time`.

Other Tips This structure works well: [base prompt] in the style of [style], [aesthetic]

Try `first person view` to improve POV results.

Try experimenting with emojis!

Don't be afraid to ask for something abstract and see what you get. (Source RunwayML)

DeepArt and Artistic Prompts

Table of Contents

Getting to Know DeepArt

In the realms of digital art and AI, DeepArt stands as a fascinating intersection that offers the promise of transforming ordinary images into masterpieces reminiscent of famous art styles. Unlike traditional filters found in photo editing applications, DeepArt adopts an innovative approach, harnessing the power of deep neural networks to interpret and reimagine photos in the vein of iconic art movements.

DeepArt's roots trace back to a deep learning technique known as style transfer. This method relies on convolutional neural networks (CNNs)—a type of artificial neural network adept at processing visual information. CNNs have been monumental in various AI-driven image tasks, such as image recognition and classification. But with style transfer, their capabilities are steered toward a more creative endeavor, extracting the stylistic essence of one image and infusing it into another.

The workings of DeepArt can be visualized as a dance between two key components, content and style. The content typically comes from the user's chosen image, say, a photograph of a serene landscape or a beloved pet. The style, on the other hand, is derived from an artwork that exhibits a distinct artistic flair—be it Van Gogh's swirling starry nights, Monet's delicate water lilies, or Picasso's abstract geometries. The challenge for DeepArt is to craft a new image that retains the content of the original photo but is cloaked in the artistic style of the reference artwork.

This is no easy feat. DeepArt's neural networks undergo rigorous training on vast collections of images. During this training phase, they learn to discern the minutiae of various styles, understanding the intricacies of brush strokes, color palettes, and textures that define different art genres. Once trained, these networks can dissect an artwork, grasp its stylistic nuances, and then apply them to another image, reshaping it while preserving its inherent content.

The results are often mesmerizing, transforming mundane snapshots into images that could easily find a place in art galleries. But it's not just about mimicking famous art styles. Users can introduce any image as a style reference, allowing for endless creative possibilities. This opens the door for custom digital art, where one could, for instance, envision their portrait in the style of a child's crayon drawing or a pattern from a cherished piece of fabric.

Beyond aesthetics, DeepArt's emergence underscores a broader cultural shift. It challenges our perceptions of creativity, raising questions about the role of machines in artistic endeavors. Historically, art has been an exclusively human domain, a reflection of our emotions, experiences, and interpretations of the world around us. DeepArt nudges us to ponder: Can machines, devoid of consciousness and emotions, truly create art? Or are they merely replicating patterns learned from their training data?

Furthermore, while DeepArt democratizes art creation, allowing anyone with a digital device to craft visually stunning images, it also sparks debates around authenticity and originality. If anyone can conjure a Van Gogh–esque masterpiece with a few clicks, what becomes of the intrinsic value of such art? Does it diminish the painstaking efforts of artists who spend years honing their craft?

In conclusion, DeepArt is emblematic of the remarkable strides made in the confluence of art and technology. While its ability to produce dazzling images is undeniable, it also serves as a mirror, reflecting societal contemplations about the evolving nature of creativity in an age dominated by algorithms. As we marvel at the digital canvases birthed by this technology, we are also prompted to delve deeper, considering the implications of AI's foray into realms once deemed uniquely human.

How Prompts Influence DeepArt

The realm of digital artistry has witnessed a significant paradigm shift with the advent of neural network–based artistic tools. Among them, DeepArt, inspired by the groundbreaking style transfer techniques, stands out. At its core, DeepArt uses prompts to infuse digital images with artistic flair, marrying content with style. But how exactly do these prompts influence the creations in DeepArt? Let's delve deeper.

The Essence of Style Transfer Before understanding prompts, one must grasp the concept of style transfer. The premise is simple yet profound: retain the content of one image while imbuing it with the style of another. Here, prompts play the role of the "style donor." They serve as a template, directing the AI on the stylistic features to capture and superimpose on the target image.

Prompts as an Artistic Compass In the vast neural landscape where DeepArt operates, prompts are akin to a compass. Without them, the

AI lacks direction. But with a clear prompt, the AI gains a defined artistic trajectory. It now has a tangible style goal to achieve, whether it's Van Gogh's swirling stars, the vibrant geometry of cubism, or the moody strokes of an impressionist sunset. The content image becomes the canvas, and the style image or prompt dictates how the AI should paint over it.

Decoding the Style DeepArt uses convolutional neural networks (CNNs) to decode the essence of the prompt. It isn't just about copying color palettes or mimicking patterns. The neural network delves deeper, identifying textures, brushstroke tendencies, depth perception, light play, and many intricate details that might elude the human eye. This comprehensive style understanding is pivotal. It ensures that the influence of the prompt on the final image is not superficial but deeply ingrained, making the outcome genuinely artistic.

Harmonious Fusion The challenge for DeepArt lies in the harmonious fusion of content and style. A direct overlay might distort the content beyond recognition or make the style seem out of place. Prompts guide this fusion process. They provide a stylistic framework within which the content must fit. The AI continually tweaks and adjusts the image, ensuring the content remains discernible but now wrapped in the artistic aura of the prompt. This harmonization process is iterative, with the AI refining its output until the balance between content integrity and style infusion is just right.

Dynamic Outcomes The beauty of using prompts with DeepArt is the dynamic range of outcomes possible with different prompts. A single content image, when paired with varying style prompts, can lead to a myriad of unique artistic creations. This variability underscores the potency of prompts. They don't just influence the final image; they redefine it. The same photograph, when run through DeepArt with a classical painting prompt, might emerge reminiscent of a Renaissance masterpiece, but when paired with a modern art prompt, it could transform into a Picasso-esque abstraction.

User's Role in Prompt Selection Another fascinating aspect is the user's role in the prompt selection process. DeepArt provides the tools, but it's the user's aesthetic preferences, encapsulated in the prompt choice, that steer the AI. In essence, the user's artistic inclinations, channeled through the prompt, coalesce with the AI's capabilities, leading to collaborative art creation.

The Limitless Horizon of Prompts While many users might opt for renowned artworks as prompts, the possibilities are endless. Hand-drawn sketches, textile patterns, natural textures, or even personal artworks can serve as prompts, allowing users to explore a vast spectrum of style outcomes. This versatility amplifies the influence of prompts in DeepArt, turning them into more than just style donors. They become co-creators, shaping, defining, and directing the AI's artistic journey.

In conclusion, prompts aren't mere inputs in the DeepArt process. They are the heart and soul of the artistic transformation, the guiding force that directs the symphony of pixels and patterns. Through prompts, DeepArt exemplifies the beautiful confluence of human creativity and machine precision, crafting digital masterpieces that resonate with both emotion and algorithmic brilliance.

Use Cases and Examples

Personalized Digital Art

Description: Users can transform their personal photographs into digital artworks that mimic the styles of iconic painters or any desired style.

Example: A family portrait could be reimagined in the style of Van Gogh's "Starry Night," infusing the image with swirling blues and golden hues.

Custom Merchandise Design

Description: Businesses can use DeepArt to create custom designs for merchandise such as T-shirts, mugs, and posters.

Example: A coffee shop might use a photograph of their signature latte and merge it with an abstract art prompt to create a unique design for their coffee mugs.

Enhanced Film and Animation

Description: Film and animation studios can utilize DeepArt to introduce distinct artistic styles into scenes or entire projects.

Example: An animation studio could take a traditionally animated scene and overlay it with a Gothic art style for a dream sequence.

Reviving Historical Photos

Description: Old black-and-white photos can be given a new life by integrating them with colorful artistic prompts, adding depth and vibrancy.

Example: A monochrome photo of a city's historical downtown can be blended with a Renaissance painting style, adding rich colors and textures and making it more engaging for modern viewers.

Album Covers and Music Promotion

Description: Musicians and bands can transform generic photos into artistic album covers or promotional materials.

Example: A band might use a basic image of their instruments but use a graffiti art prompt to produce a vibrant and street-style cover for their next album.

Digital Advertising

Description: Brands can use DeepArt to craft unique advertisements that stand out, resonating with specific artistic tastes of their target demographics.

Example: A luxury brand might merge product photos with prompts from classic paintings, highlighting the timelessness and elegance of their products.

Gaming and Virtual Reality

Description: Game developers can use artistic prompts to generate diverse in-game textures, backgrounds, or character designs.

Example: In a game set in an artistic world, each level might be influenced by a different art movement, with Impressionism, Cubism, and Surrealism influencing the design of various game levels.

Interior Design Visualization

Description: Interior designers can provide clients with unique visualizations of spaces, applying various art styles to show different moods or aesthetics.

Example: A basic 3D model of a living room could be processed with a watercolor prompt to give clients a soft, painterly visualization of a potential design.

Educational Tools for Art Classes

Description: Educators can use DeepArt to demonstrate different art styles and their influence, making art history and theory more interactive.

Example: Students could be tasked with taking a modern image, such as a cityscape, and using various prompts to transform it through the lenses of different art movements, deepening their understanding of each style.

Fashion and Textile Design

Description: Fashion designers can experiment with different artistic styles to visualize patterns, designs, or overall looks for their collections.

Example: A designer might merge a fabric's image with an Art Nouveau prompt, leading to intricate, nature-inspired patterns for their next line.

In the context of an increasingly digital world, DeepArt bridges the gap between traditional artistry and modern technology. With the power of artistic prompts, it offers endless possibilities for creators across industries. Whether for personal expression or commercial innovation, DeepArt harnesses the rich tapestry of art's history and the boundless potential of AI to craft visuals that captivate, inspire, and transcend the ordinary.

Limitations and Areas for Improvement

DeepArt, while revolutionary, has its share of constraints and potential enhancements. The integration of artistic prompts with AI models such as DeepArt has produced an innovative approach to generating artistic imagery. However, as with any emerging technology, there are notable limitations and areas awaiting improvement.

Loss of Unique Style

Concern: The risk of homogenizing art styles exists. As more users feed similar prompts or choose popular styles, there's potential for diverse art to converge toward a few trendy aesthetics.

Improvement: DeepArt could introduce random variations or encourage users to blend multiple styles, ensuring a more diverse range of outputs.

Over-Reliance on Popular Artistic Templates

Concern: Over time, users might gravitate toward universally admired artwork styles, potentially stifling creativity.

Improvement: Curating a rotating selection of lesser-known art styles or incorporating emerging artists' styles might invigorate creativity and promote lesser-known art forms.

Computational Demands

Concern: High-quality art synthesis requires significant computational resources, which might not be accessible to all users.

Improvement: Developing lightweight models or offering cloud-based processing solutions can democratize access to the technology.

Lack of Consistency

Concern: Multiple runs with the same prompt can produce varied results, which might not always align with users' expectations.

Improvement: Refining the model to offer more consistent outputs or providing users with finer control over parameters can address this.

Potential for Misuse

Concern: There's a potential risk of users appropriating styles from artists without proper acknowledgment or permission.

Improvement: Integrating mechanisms to credit original artists or even detect and prevent outright art theft can foster ethical use.

Complexity for Novice Users

Concern: For those unfamiliar with the world of art, choosing or creating effective prompts might be overwhelming.

Improvement: Introducing guided tutorials, preset prompt collections, or AI-based prompt suggestions can make the platform more user-friendly.

Artistic Authenticity

Concern: The boundary between human creativity and machine-generated art becomes blurred, raising questions about the authenticity of such art.

Improvement: Clearly marking or distinguishing AI-assisted artworks can ensure transparency and foster genuine appreciation for both human and machine contributions.

Overemphasis on Visual Aspects

Concern: Art isn't merely about aesthetics. Emotion, context, and intent play crucial roles. DeepArt might capture the visual style but miss the deeper emotional context of certain prompts.

Improvement: Research into emotion recognition and context understanding can pave the way for AI models that grasp more than just the visual cues of a prompt.

Lack of Integration with Other Media

Concern: DeepArt primarily focuses on still 2D images, leaving out possibilities with video, 3D art, or interactive media.

Improvement: Expanding the model to work with other forms of media can widen its appeal and utility.

Economic Implications for Artists

Concern: As AI-generated art becomes more prevalent, traditional artists might face challenges in monetizing their skills or finding unique value propositions.

Improvement: Platforms such as DeepArt can collaborate with artists, offering them avenues to monetize their styles, provide training data, or even teach AI-guided art courses.

In conclusion, while DeepArt and artistic prompts offer an avant-garde means to reimagine art in the digital age, it's crucial to address the limitations and continuously refine the platform. Striking a balance between technological innovation and preserving the sanctity and authenticity of art is paramount. As AI and art further intertwine, they hold the promise of ushering in an era where machines augment human creativity, rather than supplant it, leading to a richer, more diverse artistic landscape.

The Future of DeepArt

DeepArt represents a significant stride in the blending of AI and the artistic realm, but it's merely the tip of the iceberg. The possibilities are vast and hold great potential for reshaping the way we perceive and engage with art. Here's a glimpse into the likely trajectory for the future of DeepArt.

Advanced Personalization

Projection: As AI models such as DeepArt become more sophisticated, users will enjoy greater personalization, where the AI adapts to individual artistic tastes and inclinations.

Impact: This could lead to a more immersive and individualized art-creation experience, where outputs align closely with personal preferences.

Interactive Art Creation

Projection: Future iterations might evolve to allow real-time interaction, letting users adjust styles, merge influences, or toggle parameters in real time.

Impact: This will bridge the gap between static style transfer and dynamic art creation, making the process more engaging and intuitive.

Expanded Medium Support

Projection: Beyond 2D imagery, DeepArt could venture into video, 3D sculptures, and augmented reality.

Impact: Artists and creators could leverage DeepArt in multimedia projects, potentially redefining fields such as movie production, gaming, and virtual art installations.

Collaborative AI-Human Projects

Projection: Future platforms might facilitate collaboration between AI and human artists, combining the strengths of both.

Impact: Such synergy could lead to unparalleled artistic ventures, where AI handles technicalities while humans infuse emotion and context.

Emotion-Driven Art Generation

Projection: DeepArt could evolve to discern and respond to human emotions, possibly using inputs such as facial expressions, voice tonality, or physiological signals.

Impact: Art generation driven by real-time emotion detection could pave the way for therapeutic applications or personalized mood-based art experiences.

AI as an Art Tutor

Projection: Beyond generating art, DeepArt might evolve into a tutor, offering guidance, critiques, and suggestions to budding artists.

Impact: Such advancements could democratize art education, making high-quality guidance accessible to many.

Enhanced Ethical Safeguards

Projection: With concerns about art theft or undue appropriation, future versions of DeepArt might incorporate enhanced ethical safeguards.

Impact: This will promote responsible usage, ensuring artists' rights are protected, and fostering trust in the platform.

Community-Driven Evolution

Projection: Future platforms might allow the community to train and refine the AI, contributing styles, techniques, or even new algorithms.

Impact: This crowdsourced approach could ensure the tool stays relevant, diverse, and in tune with the global art community's needs.

Economic Opportunities for Artists

Projection: Artists could monetize their unique styles, allowing users to access them on DeepArt platforms, thereby offering a new revenue stream.

Impact: Such models could help artists gain wider recognition and ensure they benefit economically from the AI-art revolution.

Integration with Virtual Reality (VR)

Projection: DeepArt could merge with VR platforms, allowing users to step into AI-generated artscapes or even create in virtual 3D spaces.

Impact: This would redefine immersive art experiences, with potential applications in entertainment, education, and therapy.

The future of DeepArt and artistic prompts appears bright and boundless. As AI continues to evolve and our understanding of art deepens, platforms such as DeepArt have the potential to usher in an era of unprecedented artistic innovation. The key lies in balancing technological advancement with artistic integrity, ensuring we harness AI not just as a tool but as a collaborator in the timeless human endeavor of art creation.

22

Midjourney

Table of Contents

Introduction to Midjourney

Midjourney is an independent research lab exploring new media of thought and expanding the imaginative powers of the human species. One of their most notable projects is a Discord bot that can generate images from text prompts. This bot is currently in beta testing, but it has already gained a large following among artists, designers, and creatives of all kinds.

Midjourney is a powerful tool for creative expression, and it is still under development. The possibilities are endless, and we can't wait to see what people create with it in the future.

How to Use Midjourney

To use Midjourney, you must first create an account and join the Discord server. Once you are on the server, you can use the `/imagine` command to generate images from text prompts. For example, if you type "`/imagine` a cat sitting on a beach," the bot will generate an image of a cat sitting on a beach.

What Is Different

Midjourney offers a number of benefits for artists, designers, and creatives of all kinds. Here are just a few:

- **Creativity boost:** Midjourney can help you to spark your creativity and generate new ideas.
- **Time savings:** Midjourney can help you to create images quickly and easily, without having to spend hours sketching, painting, or rendering.
- **Versatility:** Midjourney can be used to create images in a wide range of styles, from realistic to abstract. This makes it a versatile tool for a variety of creative projects.
- **Accessibility:** Midjourney is still under development, but it is already relatively accessible to anyone with a computer and an Internet connection.

Here are some of the potential applications of Midjourney:

- **Art and design:** Midjourney can be used to create new and innovative forms of art and design. Artists can use it to generate new ideas and inspiration, and designers can use it to create unique and visually appealing products.
- **Education:** Midjourney can be used to create educational materials that are both engaging and informative. For example, teachers could use Midjourney to generate illustrations for textbooks or create interactive learning experiences.
- **Entertainment:** Midjourney can be used to create new and exciting forms of entertainment. For example, filmmakers could use Midjourney to generate concept art for new movies or TV shows, and game developers could use it to create new and immersive worlds for players to explore.

Midjourney Prompts

Prompts are the text descriptions that users provide to Midjourney to generate images. They play a vital role in the quality and creativity of the generated output. Well-crafted prompts can help users to achieve their desired results more efficiently and effectively.

Impact on Accuracy and Consistency

By providing clear and concise instructions, prompts can help Midjourney to generate images that are more accurate and consistent with the user's vision. This is especially important for professional applications, such as product design, marketing, and advertising. For example, a product designer could use prompts to generate images of new product concepts or to explore different design variations. A marketing manager could use prompts to generate images for social media posts, product packaging, or advertising campaigns.

For example, instead of simply prompting Midjourney to generate an image of a chair a user could prompt it to generate an image of a "modern wooden chair with a black leather seat and backrest." This more specific prompt would help Midjourney to generate an image that is more accurate and consistent with the user's vision.

Impact on Creativity and Innovation

Prompts can also be used to stimulate creativity and innovation. For example, users can use prompts to explore new ideas and concepts or to experiment with different artistic styles. This can lead to the development of unique and visually appealing images that would not be possible otherwise.

For example, instead of simply prompting Midjourney to generate an image of a landscape, a user could prompt it to generate an image of a "surreal landscape with floating islands and cascading waterfalls." This more creative prompt would give Midjourney more freedom to explore different ideas and to generate a more unique image.

Impact on Efficiency and Productivity

Effective prompts can help users to save time and improve their productivity. By carefully crafting their prompts before generating images, users can avoid having to make multiple attempts or having to revise their images extensively.

For example, instead of simply prompting Midjourney to generate an image of a logo, a user could prompt it to generate a "minimalist logo with a bold font and a black-and-white color palette." This more specific prompt would help Midjourney to generate an image that is more likely to meet the user's needs on the first try.

Prompts are a powerful tool that can be used to control the output of Midjourney. By carefully crafting their prompts, users can achieve more accurate, creative, and efficient results.

Midjourney is still under development, but it has the potential to revolutionize the way we create, learn, and experience the world around us. Prompts play a vital role in this process by giving users the power to shape the output of Midjourney to their specific needs and desires.

Real-World Use Cases and Examples

Midjourney is a powerful AI art generator that has been used to create stunning and creative images for a wide range of real-world use cases. Here are a few examples:

Product design: Midjourney can be used to generate images of new product concepts or to explore different design variations. For example, a product designer could use Midjourney to generate images of new furniture designs, fashion designs, or packaging designs.

Marketing: Midjourney can be used to generate images for social media posts, product packaging, or advertising campaigns. For example, a marketing manager could use Midjourney to generate images of people using a new product, images of product features and benefits, or images that evoke the desired brand image.

Web design: Midjourney can be used to generate images for website headers, banners, or landing pages. For example, a web designer could use Midjourney to generate a hero image for a new website that is both informative and visually appealing.

Video games: Midjourney can be used to generate concept art for new video games or to create textures and other assets for existing games. For example, a video game developer could use Midjourney to generate images of new character designs, environment designs, or prop designs.

Movies and TV shows: Midjourney can be used to generate concept art for new movies and TV shows or to create special effects and visual effects. For example, a filmmaker could use Midjourney to generate images of new alien creatures, futuristic worlds, or magical landscapes.

Education: Midjourney can be used to create educational materials that are both engaging and informative. For example, teachers could use Midjourney to generate illustrations for textbooks or create interactive learning experiences. For instance, a history teacher could use Midjourney to generate images of historical events, or a science teacher could use Midjournal to generate images of scientific concepts.

Research: Midjourney can be used to generate images to help researchers visualize their data and findings. For example, a medical researcher could use Midjourney to generate images of 3D models of cells or proteins, or an astronomer could use Midjourney to generate images of galaxies and nebulae.

Architecture and design: Midjourney can be used to generate images of new architectural designs or to explore different interior design options. For example, an architect could use Midjourney to generate images of different skyscraper designs, or an interior designer could use Midjourney to generate images of different living room layouts.

Music: Midjourney can be used to generate album art or music videos. For example, a musician could use Midjourney to generate an image of a surreal landscape for their album art, or a music video director could use Midjourney to generate images for different scenes in a music video.

These are just a few more examples of the many ways that Midjourney is being used in the real world. As Midjourney continues to develop and become more accessible, we can expect to see even more innovative and creative uses for this powerful AI art generator.

One of the most exciting things about Midjourney is its potential to democratize creativity. With Midjourney, anyone can create beautiful and original images, regardless of their artistic skills or experience. This opens up new possibilities for people in a wide range of fields, from education and research to marketing and advertising.

Midjourney is still under development, but it has already made a significant impact on the way we create and share art. With its powerful capabilities and its potential to democratize creativity, Midjourney is a tool that we should all be excited about.

Limitations and Areas for Improvement

Bias One of the biggest challenges facing Midjourney, and AI art generators in general, is the issue of *bias*. As mentioned before, Midjourney is trained on a massive dataset of images and text, but this dataset may contain biases. As a result, Midjourney may generate images that are biased in certain ways. For example, Midjourney may be more likely to generate images of white people than people of color, or images of men than women.

Lack of Control Another challenge facing Midjourney is the issue of **control**. Midjourney is a complex AI system, and it can be difficult to control the output. Even if you provide a very detailed prompt, Midjourney may generate an image that is different from what you had in mind. This can be frustrating for users who are trying to create specific types of images.

Speed and Cost Finally, Midjourney can be *slow and expensive*. Midjourney is still under development, and it can take several minutes or even hours to generate a single image. Additionally, access to Midjourney is currently limited to a small number of users, and it is expected to be a paid service once it is released to the public.

Despite these challenges, Midjourney is a powerful and innovative AI art generator with the potential to revolutionize the way we create and share art.

Here are a few areas where Midjourney could be improved:

Bias

One way to reduce bias in Midjourney is to use techniques such as adversarial training. In adversarial training, two neural networks are trained against each other. One neural network is trained to generate images, while the other neural network is trained to identify and flag biased images. This process helps the image-generating neural network to learn to avoid generating biased images.

Control

One way to improve control over the output of Midjourney is to develop new features that allow users to specify more details about the desired image. For example, users could be allowed to specify the desired color palette, the desired composition, or the desired style of the image.

Speed

One way to improve the speed of Midjourney is to use more powerful computing hardware. Another way to improve speed is to develop new algorithms that are more efficient at generating images.

Cost

One way to make Midjourney more affordable is to offer a variety of pricing plans. For example, Midjourney could offer a free plan for users who only need to generate a small number of images and a paid plan for users who need to generate a large number of images or who need access to more advanced features.

Despite its limitations, Midjourney is a powerful and innovative AI art generator with the potential to revolutionize the way we create and share art. As Midjourney continues to develop and improve, we can expect to see it used in even more innovative and exciting ways.

In addition to the areas for improvement listed above, I would also like to see Midjourney become more accessible to a wider range of users. Currently, Midjourney is in beta testing, and access is limited to a small number of users. I hope that Midjourney will be released to the public soon and that it will be offered at a price that is affordable to everyone.

The Future of Midjourney

As this generative AI technology continues to develop and become more accessible, we can expect to see it used in even more innovative and creative ways.

Here are some specific examples of how Midjourney could be used in the future:

Education: Midjourney could be used to create interactive learning experiences that are both engaging and informative. For example, students could use Midjourney to generate images that represent complex historical events or scientific concepts. Teachers could use Midjourney to create customized lesson plans that are tailored to the needs of their students.

Research: Midjourney could be used to help researchers visualize their data and findings in new and innovative ways. For example, scientists could use Midjourney to generate images that represent the results of their experiments. Medical researchers could use Midjourney to generate images of 3D models of cells or proteins.

Product design and marketing: Midjourney could be used to create new and innovative product designs and marketing campaigns. For example, product designers could use Midjourney to generate concepts for new products or to explore different design variations. Marketing professionals could use Midjourney to generate images for social media posts, product packaging, or advertising campaigns.

Architecture and design: Midjourney could be used to create new and innovative architectural designs and interior design options. For example, architects could use Midjourney to generate concepts for new buildings or to explore different design options for existing buildings. Interior designers could use Midjourney to generate images of different living room layouts or to explore different color palettes for a room.

Entertainment: Midjourney could be used to create new and innovative forms of entertainment. For example, filmmakers could use Midjourney to generate concept art for new movies or to create special effects and visual effects. Video game developers could use Midjourney to create new and immersive worlds for players to explore.

Personal expression and creativity: Midjourney could be used by people of all skill levels to express themselves creatively and to create beautiful and original images. For example, artists could use Midjourney to generate new inspiration for their work, or people could use Midjourney to create unique and personal gifts for their loved ones.

In addition to these specific examples, Midjourney could also have a broader impact on society by helping us to better understand ourselves and the world around us, and by providing us with new tools to solve some of the world's most pressing problems. For example, Midjourney could be used to:

Create new forms of social interaction and collaboration: Midjourney could be used to create new ways for people to connect and collaborate with each other across cultures and borders. For example, people could use Midjourney to create shared virtual spaces where they can collaborate on creative projects or simply socialize.

Help us to better understand ourselves and the world around us: Midjourney could be used to generate new perspectives on complex issues, such as climate change, poverty, and disease. This could help us to better understand these issues and to develop more effective solutions.

Solve some of the world's most pressing problems: Midjourney could be used to generate new ideas and solutions to some of the world's most pressing problems. For example, Midjourney could be used to generate new renewable energy technologies or to visualize the spread of disease.

Overall, the future of Midjourney is very bright. It is a powerful tool with the potential to make a significant positive impact on the world.

Midjourney Prompt Syntax

A midjourney prompt has up to four elements

the command `< /imagine >`

image URLs

a text prompt

parameters.

/Imagine Every midjourney prompt starts with the command `< /imagine >`. A simple prompt looks like the following:

`/imagine` [prompt A cat with wings flying in the style of Salvador Dali]

The Text Prompt The Text Prompt, or text description, is the most crucial part of a Midjourney prompt. Use it to describe the image you want Midjourney to create. We'll see that in more detail soon.

Advanced Prompts—URLs and Parameters Advanced prompts add to the other two elements and observes the following structure:

< /imagine > + < image URLs > + < text prompt > + < --parameter1 --parameter2 >

If you want Midjourney to resemble other pictures in style and content, or get consistent outputs with a series of photographs, you can insert image URLs to the prompt.

Rule: An image prompt *should always be the first element* after < /imagine >.

Lastly, the fourth component of a Midjourney prompt is the `Parameters`. It is possible to use them to change aspect ratios, quality, randomness, or even to avoid the AI inserting some specific element into the final result. Para-meters always come at the end of the prompt after < -- >. Midjourney accepts more than one parameter. You may either add a double hyphen < -- > or an em-dash (–) before the parameter.

23

Google Bard

Table of Contents

Introduction to Google Bard

Google Bard is a large language model (LLM) chatbot developed by Google AI. It is a factual language model from Google AI, trained on a massive dataset of text and code. It can generate text, translate languages, write different kinds of creative content, and answer your questions in an informative way. It is still under development, but it has learned to perform many kinds of tasks, including:

Answer Questions in a Comprehensive and Informative Way

Bard can be used to answer a wide range of questions, from simple factual questions to complex open-ended questions. For example, you could ask Bard, "What is the capital of France?" or "What are the ethical implications of artificial intelligence?" or "How do I build a birdhouse?" Bard is designed to be informative and comprehensive, and it can access and process information from the real world through Google Search. This allows it to provide answers that are both accurate and up-to-date.

Generate Different Creative Text Formats

Bard can be used to generate different creative text formats, such as poems, code, scripts, musical pieces, emails, or letters. Bard can be a powerful tool for creativity, helping you to explore new ideas and concepts or to experiment with different artistic styles. This can lead to the development of unique and visually appealing content that would not be possible otherwise. For example, you could ask Bard to write a poem about a cat or to generate a code snippet to solve a specific programming problem.

Translate Languages

Bard can be used to translate text from one language to another. This can be a useful tool for communication and collaboration across cultures and borders. For example, you could ask Bard to translate a document from English to Spanish or to translate a conversation between two people who speak different languages.

Summarize Text

Bard can be used to summarize long pieces of text into a shorter and more concise form. This can be useful for getting a quick overview of a topic or for identifying the key points of a document. For example, you could ask Bard to summarize a news article or to summarize a scientific paper.

Help with Research Tasks

Bard can be used to help with research tasks by providing summaries of relevant sources, generating new hypotheses, and identifying potential solutions to problems. Bard can be a valuable tool for researchers in all fields of study. For example, you could ask Bard to provide summaries of research papers on a particular topic, or to generate new hypotheses about a scientific phenomenon.

Improve Productivity

Bard can be used to improve productivity by automating tasks such as writing reports, generating marketing copy, and translating documents. Bard can free up your time so that you can focus on more important tasks. For example, you could ask Bard to write a report on your sales performance or to generate marketing copy for a new product launch.

Bard is still under development, but it has the potential to revolutionize the way we learn, create, and work. As it continues to learn and improve, we can expect to see even more innovative and useful applications for this powerful language model.

Prompts for Google Bard

Prompts are essential for the performance of Google Bard. They provide Bard with the information and context it needs to generate accurate, informative, and creative responses. By carefully crafting their prompts, users can achieve a wide range of benefits, including:

Increased Accuracy and Consistency

Prompts can help Bard to generate responses that are more accurate and consistent with the user's intent. This is especially important for factual tasks, such as question-answering and summarization. For example, the prompt "What is the capital of France?" will likely produce a more accurate response than the prompt "What is the capital of the world?"

Prompts can also help to improve the consistency of Bard's output, especially when generating longer or more complex responses. For example, if you ask Bard to write a blog post about a particular topic, you can use prompts to specify the desired tone, style, and format of the post. This will help to ensure that the post is consistent throughout and that it meets your specific needs.

Enhanced Creativity and Innovation

Prompts can also be used to stimulate Bard's creativity and innovation. For example, users can employ prompts to explore new ideas and concepts, to experiment with different creative text formats, or to generate new solutions to problems. This can lead to the development of unique and engaging content that would not be possible otherwise.

For example, you could ask Bard to write a poem in the style of a particular poet or to generate a code snippet to solve a complex programming problem. You could also ask Bard to generate a list of new product ideas or to brainstorm solutions to a social or environmental challenge.

Improved Control and Efficiency

Prompts can also be used to control the output of Bard and to improve efficiency. For example, users can use prompts to specify the desired output format, the desired tone of voice, or the desired length of the response. This can help users to get the desired results more quickly and efficiently.

For example, if you need Bard to write a short email to a customer, you can use a prompt to specify the desired tone and format of the email. You can also use a prompt to specify the maximum length of the email, ensuring that it is concise and to the point.

Greater Flexibility and Versatility

Prompts enable users to leverage Bard's capabilities for a wide range of tasks, including:

- **Question-answering:** Bard can be used to answer a wide range of questions, from simple factual questions to complex open-ended questions.
- **Summarization:** Bard can be used to summarize long pieces of text into a shorter and more concise form.
- **Translation:** Bard can be used to translate text from one language to another.
- **Creative writing:** Bard can be used to generate different creative text formats, such as poems, code, scripts, musical pieces, email, and letters.
- **Research assistance:** Bard can be used to help with research tasks by providing summaries of relevant sources, generating new hypotheses, and identifying potential solutions to problems.

Prompts can also be used to combine these tasks in new and innovative ways. For example, you could ask Bard to generate a summary of a research paper in the form of a poem or to translate a code snippet from one programming language to another.

Overall, prompts are a powerful tool that can be used to enhance the performance of Google Bard in a variety of ways. By carefully crafting their prompts, users can achieve more accurate, informative, creative, and efficient results.

Real-World Use Cases and Examples

Google Bard is a powerful language model with a wide range of potential real-world use cases and examples. It can be used to improve productivity and efficiency in a variety of fields, including education, research, creative writing, business, customer service, health care, and law.

Here are some specific examples of how Google Bard is being used in the real world today.

Education

Bard is being used to create personalized learning experiences for students of all ages. For example, Bard can generate interactive exercises, provide feedback on student work, and translate educational materials into different languages. This can help to ensure that all students have the opportunity to learn and succeed, regardless of their background or individual needs.

Bard is also being used to develop new educational tools and resources. For example, Bard can be used to generate educational games, simulations, and virtual reality experiences. This can help to make learning more engaging and effective for all students.

Research

Bard is being used to help researchers to generate new hypotheses, design experiments, and analyze data. For example, Bard can summarize large amounts of research literature, identify patterns in data, and generate new theories. This can help to accelerate the pace of scientific discovery and lead to new breakthroughs in a variety of fields.

Bard is also being used to develop new research tools and resources. For example, Bard can be used to generate code for data analysis, to develop new algorithms, and to write research papers. This can help researchers to be more productive and efficient in their work.

Creative Writing

Bard is being used to help writers to brainstorm new ideas, develop their characters, and improve their writing style. For example, Bard can generate different creative text formats, such as poems, stories, code, and scripts. This can help writers to overcome writer's block and to produce high-quality creative content.

Bard is also being used to develop new creative writing tools and resources. For example, Bard can be used to generate prompts for writers, to provide feedback on writing samples, and to translate creative writing into different languages. This can help writers to be more creative and productive in their work.

Business

Bard is being used to improve productivity and efficiency in a wide range of business settings. For example, Bard can automate tasks such as writing reports, generating marketing copy, and translating documents. This can free up employees' time so that they can focus on more important tasks.

Bard is also being used to develop new business tools and resources. For example, Bard can be used to generate customer insights, to develop new products and services, and to create marketing campaigns. This can help businesses to be more competitive and successful in the marketplace.

Customer Service

Bard is being used to provide customer support and answer customer questions in a timely and accurate manner. For example, Bard can generate FAQs, chat with customers online, and resolve customer issues. This can help to improve the customer experience and reduce the workload on customer service representatives.

Bard is also being used to develop new customer service tools and resources. For example, Bard can be used to generate knowledge bases, to develop chatbots, and to translate customer support materials into different languages. This can help businesses to provide better customer service to their customers around the world.

Health Care

Bard is being used to improve the quality of health care delivery. For example, Bard can help doctors to diagnose diseases, develop treatment plans, and communicate with patients. Bard can also be used to generate educational materials for patients and their families.

Bard is also being used to develop new health care tools and resources. For example, Bard can be used to develop AI-powered diagnostic tools, to generate personalized treatment plans, and to translate health care materials into different languages. This can help to make health care more accessible and affordable for people around the world.

Law

Bard is being used to improve the efficiency and effectiveness of legal research and writing. For example, Bard can generate legal summaries, identify relevant case law, and draft legal documents. This can help lawyers to be more productive and efficient in their work.

Bard is also being used to develop new legal tools and resources. For example, Bard can be used to develop AI-powered legal research tools, to generate personalized legal advice, and to translate legal documents into different languages. This can help to make legal services more accessible and affordable for people around the world.

These are just a few examples of the many ways that Google Bard is being used in the real world today. As Bard continues to develop and improve, we can expect to see even more innovative and powerful applications.

Limitations and Areas for Improvement

As with any generative AI platform and some of which we have covered in this book, Bard has limitations. Some of these are:

Accuracy and Reliability

Bard is trained on a massive dataset of text and code, but it is still not perfect. It can sometimes generate inaccurate or misleading information, especially for complex or open-ended questions or for topics that are rapidly evolving.

Bias and Fairness

Bard's responses may be biased against certain groups or individuals, as it is trained on a dataset that reflects biases and stereotypes that exist in the real world.

Creativity

Bard can generate creative text formats, but its creativity is still limited. It can sometimes produce repetitive or unoriginal content.

Context

Bard can sometimes struggle to understand and respond to context-rich prompts. This is because Bard is trained on a massive dataset of text and code, but it does not have the same understanding of the world as a human does.

Areas for Improvement

Accuracy and reliability: Google AI is actively working to improve Bard's accuracy and reliability by training it on more data and developing new techniques for detecting and correcting errors.

Bias and fairness: Google AI is working to address bias and fairness in Bard by developing new training methods and by working with experts to identify and mitigate potential biases.

Creativity: Google AI is working to improve Bard's creativity by developing new training methods and by exposing Bard to a wider range of creative content.

Context: Google AI is working to improve Bard's context awareness by developing new training methods and by providing Bard with more information about the world, such as commonsense knowledge and factual knowledge.

Additional Thoughts

In addition to the limitations and areas for improvement listed above, it is important to note that Bard is still under development. This means that its capabilities are constantly expanding and changing. It is also important to remember that Bard is a machine learning model, and it is not a sentient being. It does not have the same understanding of the world as a human does, and it can sometimes make mistakes.

It is important to use Bard responsibly and to be aware of its limitations. When using Bard, it is important to carefully review its responses and to be critical of the information it provides. If you are unsure about the accuracy or reliability of a response, it is best to consult with a human expert.

Overall, Google Bard is a powerful and versatile language model with the potential to revolutionize the way we learn, create, and work. However, it is important to be aware of its limitations and to use it responsibly.

The Future of Google Bard

Google Bard has the potential to revolutionize the way we learn, create, work, and live. It can be used to:

Improve our understanding of ourselves and the world around us: Bard can analyze large amounts of data from a variety of sources, including scientific research papers, news articles, and social media posts. This can help us to identify patterns and trends and to generate new insights into the world around us. Bard can also be used to analyze our personality traits, cognitive biases, and emotional responses. This can help us to better understand our strengths and weaknesses, and to make better decisions in our lives.

Solve complex problems: Bard can brainstorm potential solutions to problems and evaluate the pros and cons of different solutions. Bard can also be used to develop and implement solutions to problems. For example, Bard could be used to develop new technologies to help us to solve problems such as climate change and poverty. Bard could also be used to create new products and services that improve our health, education, and well-being.

Create new art and entertainment: Bard can generate new ideas for poems, stories, movies, and video games. Bard can also be used to create new forms of art and entertainment that we have never seen before. For example, Bard could be used to generate personalized music playlists or to create interactive stories that adapt to the reader's choices.

Improve the quality of life for people around the world: Bard can be used to develop new technologies and services that can improve the lives of people in many different ways. For example, Bard could be used to develop new medical treatments, or to create new educational resources. Bard could also be used to improve communication and collaboration between people from different cultures and backgrounds.

Overall, the future of Google Bard is very bright. It has the potential to make a significant positive impact on the world. As Bard continues to develop and improve, we can expect to see even more innovative and powerful applications.

Here are some specific examples of how Google Bard could be used in the future:

- A student could use Bard to generate a personalized study plan for an upcoming exam. Bard could identify the student's strengths and weaknesses and then generate a plan that focuses on the areas where the student needs the most help.
- A scientist could use Bard to generate new hypotheses about a particular research topic. Bard could analyze a large body of research papers on the topic and then identify patterns and trends that the scientist may not have noticed.

- A writer could use Bard to come up with new ideas for stories or articles. Bard could generate a list of potential plot points or topics, and then the writer could explore those ideas further.
- A businessperson could use Bard to develop a new marketing campaign for their product or service. Bard could analyze data from social media and other sources to identify the target audience's needs and interests. Bard could then generate a marketing campaign that is tailored to those needs and interests.
- A doctor could use Bard to help diagnose a patient's illness. Bard could access a vast database of medical information and then use that information to identify the most likely causes of the patient's symptoms.

These are just a few examples of the many ways that Google Bard could be used in the future. As Bard continues to develop and improve, we can expect to see even more innovative and powerful applications.

24

Deepfaking with DeepFaceLab

Table of Contents

Introduction to DeepFaceLab

In the vast landscape of artificial intelligence, certain applications capture public attention due to their profound impact on digital content creation, ethical considerations, and potential use cases, both benign and malicious. DeepFaceLab is one such tool, playing a pivotal role in the creation of deepfakes, videos where faces are swapped or manipulated using machine learning techniques.

DeepFaceLab, by definition, is an open-source software that harnesses deep learning methodologies to superimpose the facial features of one individual onto another in videos. Powered by sophisticated neural networks, this software has democratized the deepfake creation process, allowing both professionals and enthusiasts to produce highly realistic video manipulations.

Deepfakes, a portmanteau of "deep learning" and "fake," surfaced as a technological novelty but quickly raised concerns due to their potential misuse. Their history can be traced back to research initiatives in universities and institutions aiming to advance facial recognition, modeling, and manipulation techniques. As the technology matured, applications such as DeepFaceLab emerged, providing the general public with the capability to create deepfakes with relative ease.

The functionality of DeepFaceLab is rooted in a series of machine learning models, specifically designed for facial recognition, alignment, and manipulation. These models are trained on vast datasets, learning to identify facial landmarks, textures, lighting conditions, and expressions. By doing so, the software can seamlessly blend the facial features of one individual onto a target video, ensuring the final output maintains natural motion, lighting, and emotion.

One of the primary reasons for DeepFaceLab's prominence in the deepfake community is its user-friendliness. Unlike some of its contemporaries, DeepFaceLab offers an intuitive interface, simplifying the traditionally complex process of video manipulation. Users are guided through a series of steps, from selecting source and target videos, aligning faces, training the model, and finally, rendering the manipulated video. This step-by-step process, combined with comprehensive tutorials available online, means even those with limited technical knowledge can experiment with the software.

However, it's essential to understand the broader implications of tools such as DeepFaceLab. As deepfakes become more prevalent, discerning reality from fabrication in digital media becomes increasingly challenging. Authentic videos can be dismissed as deepfakes, and manipulated content can be portrayed as genuine, raising significant concerns in contexts such as journalism, political campaigns, and personal privacy.

Moreover, the ethical boundaries of deepfake technology are still being defined. Is it right to use someone's likeness without their consent? What about creating videos that rewrite history or spread false information? While DeepFaceLab serves as a tool, its application's ethics lies in the hands of the user.

In summary, DeepFaceLab stands at the intersection of technological advancement and ethical quandary. As a pioneering tool in the realm of deepfakes, it showcases the prowess of modern AI in content creation. Yet, with great power comes great responsibility. As we continue to grapple with the potential and pitfalls of such technology, it's imperative to approach it with a discerning eye and a keen sense of ethics.

How Prompts Influence DeepFaceLab

Prompts play a critical role in shaping the outcome of deepfakes created using DeepFaceLab. Understanding the influence of prompts can help users make more informed decisions when generating deepfakes, ensuring desired outcomes while mitigating potential misuse. Here, we explore the various ways prompts affect the deepfake generation process.

Guiding the Model

Purpose definition: Prompts provide the model with a clear directive, informing it about the specific transformation required, whether it's swapping a celebrity's face onto a movie character or simulating a historical figure's speech.

Refinement: A well-structured prompt can guide the model to produce more refined and accurate results, ensuring that the deepfake meets user expectations.

Controlling Aesthetics

Facial features emphasis: Prompts can emphasize certain facial features, such as enhancing smile lines or emphasizing eye color, resulting in a more pronounced or subtle effect in the final deepfake.

Mood and expression: Through prompts, users can guide the model to focus on specific facial expressions or moods, for instance, ensuring a deepfaked individual looks surprised or pensive.

Optimizing for Context

Background consistency: Prompts can help maintain consistency with the background or surrounding environment. For instance, ensuring that the lighting on the deepfaked face matches the scene's overall lighting.

Interactive elements: If a scene involves interaction, such as a person touching their face, prompts can help the model account for such interactions, ensuring the deepfake remains convincing.

Fine-Tuning for Authenticity

Voice synchronization: When combining face-swapping with voice deepfakes, prompts can instruct the model to ensure that facial movements are in sync with the spoken words.

Physical dynamics: The model can be prompted to consider physical dynamics such as hair movement, ensuring that elements such as wind or motion don't disrupt the deepfake's realism.

Ethical Boundaries

Avoiding misrepresentation: Prompts can be designed to include watermarks or visible signs indicating that a video is deepfaked. This can help prevent unintentional misinformation or deceit.

Informed consent: By integrating prompts that ensure facial swaps only from consensual datasets, users can maintain ethical standards in their deepfake projects.

Creativity and Experimentation

Artistic ventures: Prompts allow for creative experimentation, enabling artists and enthusiasts to explore novel visual narratives, from placing historical figures in modern settings to envisioning cross-genre movie crossovers.

Unique scenarios: With the right prompts, users can instruct DeepFaceLab to generate unconventional or whimsical results, such as combining features from multiple faces or generating fantasy-inspired appearances.

Optimizing for Different Platforms

Resolution and quality: Depending on where the deepfake will be showcased, prompts can be tailored to optimize for specific resolutions, whether it's for HD television screens or social media platforms.

File size and format: Users can prompt the model to produce deepfakes in specific file sizes or formats, catering to platform-specific requirements.

Addressing Limitations

Data shortages: In scenarios with limited source footage, prompts can guide the model to fill in gaps, perhaps by referencing similar faces or extrapolating from available data.

Error corrections: If users spot anomalies in initial results, prompts can instruct the model to address those specific issues in subsequent attempts.

Safety and Security

Watermarking: To ensure that deepfakes are identifiable, prompts can include directives for the model to subtly watermark or brand the generated content, indicating its artificially generated nature.

Feedback Loop

Iterative improvements: Prompts can be adjusted based on feedback, creating an iterative process where users refine their instructions based on initial outcomes, progressively improving the quality of generated deepfakes.

While prompts significantly influence DeepFaceLab's output, users must approach the technology with an understanding of its potential societal impacts. Striking a balance between creativity and ethical considerations is paramount.

Practical Examples and Use Cases

Deepfaking, a cutting-edge technology that uses neural networks to superimpose one face onto another in videos, has sparked both fascination and concern in equal measure. At the forefront of this technology is DeepFaceLab, a tool that has made deepfaking more accessible than ever before. Here, we explore practical examples and use cases where DeepFaceLab's capabilities have been harnessed.

Film and Entertainment

Resurrecting actors: DeepFaceLab has been used in the film industry to bring deceased actors back to life for specific roles, allowing them to "appear" posthumously. For instance, an actor who passed away during the filming process can be seamlessly integrated into the remaining scenes.

Stunt doubles: Instead of placing actors in potentially dangerous situations, their faces can be superimposed onto stunt doubles, ensuring safety without compromising on authenticity.

Age regression/progression: Instead of hiring multiple actors to play the same character at different ages, filmmakers can use DeepFaceLab to digitally age or de-age an actor.

Education and Research

Historical recreations: DeepFaceLab can breathe life into historical figures by superimposing their faces onto actors in educational videos. This provides a more immersive learning experience for students studying history.

Psychological studies: Researchers can use deepfakes to study human perception, biases, or reactions to altered visual stimuli in controlled settings.

Art and Digital Media

Music videos: Artists have experimented with deepfakes in their music videos, creating unique visuals and narratives that captivate their audience.

Digital art installations: Modern artists use deepfakes as a medium to explore themes of identity, reality, and digital manipulation, leading to thought-provoking exhibitions.

Gaming

Character realism: Game developers can use DeepFaceLab to create lifelike characters by overlaying real faces onto digital avatars, enhancing the immersive experience for players.

Historical games: Deepfakes allow for the accurate representation of historical figures in games, adding an element of authenticity.

Forensics and Law

Training and simulations: Deepfakes can be used to create realistic training videos for law enforcement, helping them understand and counteract deepfake-related crimes.

Legal reenactments: In legal settings, deepfakes can help recreate scenes or events, providing visual aids during court proceedings.

Advertising

Celebrity endorsements: Brands can use deepfakes to simulate celebrity endorsements, especially if getting the actual celebrity is logistically challenging or expensive. However, this raises ethical concerns, and consent is paramount.

Global campaigns: Advertisers can adapt a single advertisement for multiple global markets by changing the faces of actors to resonate with different demographics.

Personal Projects

Ancestry exploration: Individuals can use DeepFaceLab to superimpose their faces onto old family photos, creating a bridge between generations and visualizing family resemblances.

Fictional scenarios: Hobbyists have used deepfakes to create fictional crossovers, such as placing modern actors in classic films or vice versa.

Satire and Parody

Political commentary: Satirists have employed deepfakes to create humorous or critical videos of political figures, aiming to provide commentary or spark debate.

Internet memes: The meme culture has embraced deepfakes, leading to the viral spread of humorous content.

Dubbing and Language Translation

Lip synchronization: When translating films or shows into different languages, deepfakes can adjust actors' lip movements to match the dubbed dialogue, creating a seamless viewing experience.

Journalism and Documentaries

Recreating events: Journalists can use deepfakes to recreate events or scenarios when actual footage is unavailable or too graphic to display.

While these practical examples showcase the potential of DeepFaceLab, it's crucial to tread carefully. Deepfakes, especially when used maliciously, can spread misinformation, harm reputations, and even pose national security threats. As with all powerful technologies, the ethical implications must be weighed alongside the innovative possibilities.

Ethical Considerations and Limitations

Deepfaking, especially with tools as powerful as DeepFaceLab, comes with profound ethical ramifications and limitations. The blend of technology and human intent opens avenues for both revolutionary creative applications and potentially malicious uses. Here's an in-depth look at the intertwined ethics and constraints.

Misinformation and Disinformation

Potential for abuse: Deepfakes can be weaponized to spread false information, potentially damaging reputations or swaying public opinion.

Election meddling: In politically charged environments, doctored videos of politicians or public figures can be used to deceive voters.

Consent and Privacy

Unauthorized usage: Creating deepfakes of individuals without their explicit consent infringes on personal rights and can lead to undesired public exposure.

Invasion of privacy: The potential to manipulate personal videos or images can make anyone a target, leading to breaches of privacy.

Monetary and Economic Implications

Financial markets: False videos related to companies, CEOs, or economic indicators can influence stock markets and investor sentiments.

Entertainment industry: Unsanctioned use of a celebrity's likeness can circumvent contract agreements or undermine genuine content.

Psychological and Emotional Effects

Trust erosion: As deepfakes become more prevalent, the public may begin to distrust genuine video content, leading to widespread skepticism.

Targeted harassment: Deepfake technology can be used for personal vendettas, creating harmful content to shame, embarrass, or harass individuals.

Legal Ambiguities

Undefined laws: Many jurisdictions lack comprehensive laws on deepfakes, leading to gray areas in content creation, distribution, and accountability.

Proof and prosecution: Authenticating videos in legal cases becomes challenging, potentially affecting justice outcomes.

Quality and Detection

Imperfections: While DeepFaceLab is advanced, it can still produce artifacts or glitches that betray a video's manipulated nature.

Race against detection: As deepfake creation tools evolve, so do detection tools. However, there's a constant cat-and-mouse game between creation and detection, and sometimes, malicious deepfakes can go undetected.

Hardware and Software Limitations

Processing power: Creating high-quality deepfakes requires substantial computational resources, which might not be accessible to everyone.

Training data: The accuracy and believability of deepfakes largely depend on the quality and quantity of training data. Insufficient or biased data can lead to unconvincing results.

Societal and Cultural Impacts

Normalization of distrust: The widespread use and knowledge of deepfakes can lead society toward a general distrust of digital media, affecting interpersonal relationships and community dynamics.

Misrepresentation of history: Historical figures or events could be falsely portrayed, leading to misconceptions and altered perspectives on factual events.

Commercial Misuse

Branding and reputation: Companies can be misrepresented, leading to brand damage or undesired public relations incidents.

Advertising: False endorsements or negative portrayals can skew product perceptions and influence consumer behavior.

Educational and Awareness Challenges

General awareness: The general public might be unaware of deepfake capabilities, making them more susceptible to deception.

Digital literacy: Ensuring that digital users, especially in educational settings, are aware of and can identify deepfakes is an ongoing challenge.

Addressing these ethical considerations and limitations necessitates a collaborative approach. Stakeholders, including tech developers, legislators, educators, and the general public, must be engaged in discussions and actions to harness the positive potential of DeepFaceLab while curtailing its negative consequences. As with any revolutionary technology, understanding, responsibility, and proactive measures are key to beneficial integration into society.

The Future of DeepFaceLab

The realm of deepfaking, especially with tools such as DeepFaceLab at the forefront, is evolving at an unparalleled pace. As technology continues its relentless march, DeepFaceLab and similar tools will undoubtedly mature, bringing about a future filled with both immense potential and daunting challenges. Let's delve into what lies ahead.

Advanced Realism

The primary goal of deepfaking tools such as DeepFaceLab is to increase the realism of generated content. We can anticipate improvements in:

Resolution: As computational capabilities advance, higher resolution deepfakes will become commonplace.

Micro-expressions: Capturing nuanced facial expressions will lead to more lifelike deepfakes.

Voice synchronization: Merging facial manipulation with convincing voice synthesis can lead to entirely artificial yet realistic video content.

User Accessibility

Simplified interfaces: Future versions of DeepFaceLab may prioritize user-friendly interfaces, making the technology accessible to non-experts.

Mobile platforms: As smartphones become more powerful, we might see mobile-compatible versions or lighter deepfaking apps inspired by DeepFaceLab.

Integration with Other Technologies

Virtual reality (VR) and augmented reality (AR): DeepFaceLab's capabilities could extend to VR and AR, creating immersive environments featuring real and artificial entities.

Gaming: Game developers could use advanced deepfaking tools to create characters based on real-world people or blend real actors seamlessly into virtual worlds.

Automated Content Creation

Filmmaking: Directors might use DeepFaceLab to simulate scenes before actual shoots or even replace actors in certain scenes, potentially reshaping the world of cinema.

Ad personalization: Businesses could generate personalized ads by integrating user's likenesses or preferences using deepfaking technologies.

Educational and Training Modules

Historical reenactments: Deepfaking could recreate historical figures for educational content, allowing students to "meet" these figures virtually.

Corporate training: Companies might use deepfaked scenarios for training, creating situations that look real but are entirely simulated.

Counter-Deepfaking Technologies

Detection tools: As deepfaking evolves, so will the need for detection. The future will see advanced tools capable of identifying even the most sophisticated deepfakes.

Digital watermarking: Content could be embedded with imperceptible watermarks, indicating whether it's been altered or is genuine.

Ethical and Regulatory Framework

Guidelines for use: As society becomes more aware of deepfaking's implications, there could be established guidelines or best practices for ethical use.

Legislation: Governments may introduce stringent regulations, especially for malicious use or unauthorized deepfaking of individuals.

Open-Source versus Proprietary Battles

Collaborative advancements: The open-source nature of tools such as DeepFaceLab means that developers worldwide can contribute, leading to rapid advancements.

Proprietary tools: However, as the commercial potential of deepfaking becomes clear, we might see proprietary tools emerging, offering exclusive features or capabilities.

Public Perception and Trust

Shift in trust: The ubiquity of deepfakes might lead to a significant shift in how the public perceives digital content, leading to increased skepticism.

Media literacy: Education systems might prioritize media literacy, ensuring future generations can discern genuine content from manipulated media.

Security and Verification Systems

Biometric systems: While deepfakes could challenge current facial recognition systems, it could also propel the development of advanced biometric verification tools resilient to deepfakes.

Blockchain for media: Blockchain technology could be used to verify the authenticity of media, creating immutable records for genuine content.

The future of DeepFaceLab and the broader realm of deepfaking is multifaceted. The technology's convergence with other fields and its potential applications are vast. However, this future isn't devoid of challenges. Balancing innovation with ethical use, ensuring public trust, and navigating the complex regulatory landscape will be paramount. As with most disruptive technologies, the key will lie in informed use, proactive adaptation, and societal preparedness.

25

Image Editing with DeepArt Effects

Table of Contents

Getting to Know DeepArt Effects

In today's digitally driven world, the intersection of art and technology has given rise to tools that can emulate and even enhance human artistic creativity. Among these tools, DeepArt Effects stands out as a remarkable convergence of machine learning and digital artistry. A fascinating blend of the past and the future, DeepArt Effects merges traditional artistic aesthetics with cutting-edge technology to create something truly novel for the modern user.

At its core, DeepArt Effects is an AI-powered image processing application. Unlike traditional photo editing tools, which primarily rely on filters, overlays, and manual adjustments, DeepArt Effects uses deep learning models to reinterpret and transform photographs into distinctive works of art. The user's chosen style, whether it's reminiscent of Van Gogh's swirling starry nights or the abstract cubism of Picasso, is not just overlaid on the image. Instead, the application delves deep into the nuances of the photo, analyzing its content and context, and then recreates it from scratch, ensuring the artistic style melds seamlessly with the original content.

The success and allure of DeepArt Effects can be attributed to its underlying technology: neural style transfer. This process is a feat of neural network architectures where two images—the content image (the user's photo) and the style image (a piece of artwork)—are input into the model. The AI then seeks to generate a new image, preserving the content of the original photo but with the artistic style of the artwork. This involves intricate computations, parsing through layers of the neural network to extract content features from one image and style features from the other. The final artwork is an AI's interpretation, a balance between the chosen style and the intrinsic details of the uploaded photo.

The emergence of platforms such as DeepArt Effects is emblematic of a broader trend in the digital world. As machine learning models become more sophisticated, they're increasingly encroaching into domains that were once considered the exclusive realm of human creativity. The democratization of art, facilitated by such tools, means that one no longer needs to be an accomplished artist or even have a rudimentary understanding of art techniques to create something visually appealing. With a few clicks, an ordinary photograph can be transformed into an artwork worthy of a digital gallery.

However, while the results are undeniably impressive, they also spark some contemplation about the nature of creativity. Does using DeepArt Effects diminish the value of the final artwork since the creation process is primarily automated? Or does the user's choice of image, style, and their intent constitute a form of artistic expression in itself? These questions, central to the broader debate about AI and creativity, don't have straightforward answers. But they underscore the transformative impact of technology on fields that were traditionally human centric.

Another remarkable aspect of DeepArt Effects is its accessibility. Art, in many forms, has often been viewed as exclusive—whether due to required skills, tools, or platforms. However, applications such as DeepArt Effects make artistic expression more inclusive. They act as gateways, allowing individuals who might not have engaged with art to explore, appreciate, and even create. For many, it serves as an introduction to the vast world of art styles, artists, and techniques.

In conclusion, DeepArt Effects represents the exciting potential and challenges of integrating AI into creative fields. While it offers a novel way for individuals to engage with art, it also pushes us to reconsider and redefine the boundaries of creativity. As technology continues to evolve, platforms such as DeepArt Effects will undoubtedly play a pivotal role in shaping the landscape of digital artistry, blurring the lines between man, machine, and canvas.

The Role of Prompts in DeepArt Effects

The technological symphony between AI and artistic endeavors has taken the digital realm by storm, with platforms such as DeepArt Effects leading the vanguard. While the fusion of neural networks and artistry is itself a marvel, an essential yet often overlooked component that makes this alchemy possible is the concept of prompts. Prompts, in the context of DeepArt Effects, act as guiding lights, instructing the AI on how to mold the artistic transformation of images.

User Input as Prompts Every action on DeepArt Effects, from choosing a particular art style to deciding the intensity of the transformation, serves as a prompt for the underlying AI model. When a user selects Van Gogh's "Starry Night" style, they're essentially providing a directive to the AI, instructing it to reinterpret the uploaded image with the swirling, vivid characteristics of Van Gogh's masterpiece.

Guiding the Style Transfer DeepArt Effects is underpinned by neural style transfer techniques. Here, prompts become critical. The *style image* (e.g. a famous artwork) serves as a prompt, guiding the AI on the style attributes to extract and imbue onto the user's photo. Similarly, the *content image* (user's photo) instructs the AI on the content to retain. The dance between these two prompts results in the final artistic rendition.

Custom Style Prompts Some versions of style transfer tools, including platforms such as DeepArt Effects, allow users to upload custom styles. In such scenarios, the custom artwork acts as a prompt. The AI then grapples with deciphering and applying this new, unfamiliar style onto the user's image, showcasing the model's adaptability and the significance of diverse prompts.

Intensity and Refinement Prompts aren't limited to style and content alone. Tools might offer sliders or options to adjust the intensity of the transformation. Such selections guide the AI, instructing it on the depth of style immersion. A low-intensity prompt might result in subtle artistic hints, while a high-intensity prompt could lead to a complete stylistic overhaul of the original image.

Iterative Feedback as Prompts For users unsatisfied with the initial result, their modifications and adjustments act as iterative prompts. By reediting, users provide feedback to the model, refining the output. This cyclical prompting allows for a more tailored artistic creation, where AI and user continually refine the artwork based on mutual feedback.

Limitations and Potential Misinterpretations Like any other model, the AI behind DeepArt Effects isn't perfect. There can be times when the AI misinterprets a prompt or fails to capture the nuances of a style or image content. These occasional missteps underscore the importance of clear and effective prompts. It also highlights an area where the interaction between users and AI can be improved, ensuring that the AI better understands and responds to prompts.

Ethical Implications The prominence of prompts in guiding AI-driven artistry brings up ethical considerations. As users supply prompts, there's a collaborative element in the creation process. Who then gets credit for the artwork? Is it the user who provided the guiding prompts, or is it the AI that executed the transformation? These questions are emblematic of the broader discourse around AI and creativity.

In essence, prompts serve as the bridge between human intent and AI-driven artistic transformation in platforms such as DeepArt Effects. They provide direction, ensuring that the AI's vast capabilities are channeled to align with the user's vision. As the technology behind AI-driven artistry evolves, the nuances and complexities of prompts will undoubtedly come into sharper focus. Their role will be pivotal in ensuring that the AI not only creates art but does so in a way that resonates, reflects, and respects human intent and vision.

Showcasing Real-World Scenarios

In recent years, digital artistry tools, such as DeepArt Effects, have radically transformed the way we view and modify images, bringing intricate art styles within the grasp of everyday users. As these platforms become increasingly sophisticated and accessible, a variety of real-world applications emerge, extending far beyond simple image editing. Let's delve into some of the noteworthy scenarios where DeepArt Effects has made its mark.

Personal Digital Portfolios and Social Media

Photographers and enthusiasts: Professionals and hobbyists alike are using DeepArt Effects to infuse new life into their photographs. Whether it's applying Van Gogh's signature swirls to a sunset snap or imparting the abstract features of Picasso to a portrait, the platform offers a fresh lens through which to view common images.

Influencers and bloggers: To stand out in the crowded landscape of social media, many influencers are turning to unique filters and artistic edits. By transforming everyday moments into pieces reminiscent of iconic artworks, they capture their audience's attention, making their content memorable.

Fashion and Textile Industry

Custom designs: Fashion designers are leveraging the tool to iterate novel patterns and designs for clothing. Inputting plain fabrics into DeepArt Effects and overlaying them with specific artistic styles can result in unique, wearable art pieces.

Print and graphic tees: Graphic designers are employing the platform to craft distinctive designs for apparel, particularly for T-shirts, yielding products that are both fashionable and artistic.

Entertainment and Multimedia

Music album art: Musicians and bands are exploring DeepArt Effects to create evocative album covers. By blending their photographs with certain artistic styles, they can mirror the mood or theme of their music.

Film and animation: In some indie projects, filmmakers and animators use tools such as DeepArt Effects to add artistic flairs or sequences, merging classical art vibes with modern storytelling.

Advertising and Branding

Ad campaigns: In a bid to be innovative, advertisers are implementing DeepArt's unique artistic transformations to design ads that resonate with viewers. A product shot, when combined with a specific art style, can evoke certain emotions or nostalgia, amplifying the message's impact.

Logo and brand identity: Companies looking to refresh or reinvent their brand identity are experimenting with the platform, applying classic art styles to modern logos to find a balance between tradition and contemporaneity.

Education and Research

Teaching art history: Educators in the realm of art and history are utilizing DeepArt Effects to give students a hands-on experience. By allowing students to transform their own images into the styles of famous artists, the tool provides an engaging way to learn about historical art movements and techniques.

Study of art styles: Researchers, while studying various art forms, can utilize DeepArt to simulate and better understand the nuances of particular artistic techniques, comparing AI-generated art with original masterpieces.

Home Decor and Interior Design

Custom artworks: Homeowners and interior designers are employing the platform to produce custom art pieces tailored to specific spaces. A family photograph, when rendered in the style of Monet or Matisse, can become a personalized centerpiece for a living room.

Themed spaces: For spaces themed around specific eras or art movements, designers are transforming various elements—from wall hangings to furniture textures—using DeepArt Effects to ensure thematic consistency.

Digital and Physical Art Exhibits

Art installations: Modern artists are integrating DeepArt-generated pieces into their exhibits, blurring the lines between non-digital and digital art. Some even base their entire collections around AI-generated art, prompting discussions about technology's role in the creative process.

Interactive exhibits: Museums and galleries are setting up interactive kiosks where visitors can transform their images using DeepArt, making the artistic experience personal and immersive.

The versatility and ease of use of DeepArt Effects have carved out niches in varied sectors, revolutionizing the way we perceive and interact with art. Whether for professional use or personal exploration, the melding of AI-driven artistry into real-world scenarios underscores the burgeoning symbiosis between technology and creativity.

Strengths and Weaknesses

Image editing, in its modern form, has seamlessly integrated with AI. DeepArt Effects is a striking example, melding deep learning techniques with artistry to transform mundane images into artistic masterpieces. As with any technology, it comes with a set of strengths and weaknesses that define its use and potential impact on the broader art and design landscape.

Strengths

Seamless Integration of Classic Art Styles DeepArt Effects possesses the capacity to recreate the nuanced brushstrokes and styles of legendary artists such as Van Gogh or Monet. This offers users the unique opportunity to reimagine their photographs within the confines of these timeless styles.

User-Friendly Interface For many, mastering professional-grade image editing software can be intimidating. DeepArt Effects provides an intuitive interface, ensuring that even novices can produce artwork with minimal effort.

Rapid Transformation Leveraging the computational prowess of deep neural networks, the platform can process and deliver high-quality artistic renditions in a matter of minutes. This efficiency is particularly advantageous for professionals working under tight deadlines.

Customization Beyond predefined styles, DeepArt Effects allows users to create custom templates by feeding in their favorite artworks. This promotes a sense of creative freedom, enabling unique and personalized art transformations.

Cost-Effective Art Creation Hiring professional artists or designers might be beyond the budget of many, especially for one-off projects or personal endeavors. With DeepArt Effects, users can achieve professional-grade results at a fraction of the cost.

Weaknesses

Over-Reliance on Technology While DeepArt Effects can replicate artistic styles, it might inadvertently promote an over-reliance on technology. Traditional art skills, including understanding color theory, composition, and other nuances, could become undervalued.

Lack of True Originality Since the platform bases its transformations on existing art styles, the end product, though beautiful, may lack the originality and soul inherent in manually crafted artwork.

Uniformity in Output If multiple users apply the same preset to similar images, the results can be eerily similar. This uniformity can lead to a saturation of a particular style in the market, especially if it becomes trendy.

Ethical Concerns There's a fine line between inspiration and imitation. Using DeepArt Effects to generate commercial artwork could raise ethical issues, especially if the resultant piece closely mirrors a copyrighted artwork.

Limited to 2D Media While DeepArt Effects excels in transforming 2D images, it doesn't cater to 3D modeling or animation. This limits its application in industries such as gaming, film production, or virtual reality.

Dependence on Quality of Input The platform's output heavily depends on the quality of the input image. A low-resolution or poorly lit image might not yield the desired artistic effect, potentially necessitating manual touch-ups.

Potential for Misuse The ease with which DeepArt Effects can transform images introduces the risk of misuse. Users could modify copyrighted images, claiming them as their own, leading to potential legal disputes.

In conclusion, DeepArt Effects has undeniably revolutionized the realm of image editing by democratizing access to classic art styles. Its strengths lie in its ability to provide rapid, cost-effective, and user-friendly art transformations. However, it's essential to approach its use judiciously, respecting the boundaries of originality and ethics. As the technology evolves, there's potential for many of its limitations to be addressed, but a mindful balance between manual artistry and AI-driven automation remains crucial.

Future Trajectories of DeepArt Effects

The fusion of AI and image editing, as epitomized by DeepArt Effects, holds transformative promise for the realm of digital artistry. As technology continues to advance, it is intriguing to envision the potential trajectories for platforms such as DeepArt Effects, especially in a world that increasingly values digitization, personalization, and interactivity.

Evolution into 3D and Virtual Realms Though DeepArt Effects currently excels in 2D image transformations, the future may see an expansion into the 3D space. As virtual reality (VR) and augmented reality (AR) gain traction, there will be a growing demand for tools that can render real-world environments in artistic styles in real time. Imagine walking through a VR cityscape transformed into a Van Gogh–esque world.

Enhanced Personalization through Machine Learning While DeepArt Effects already offers customization to some extent, future iterations might use machine learning to understand individual user preferences better. This would mean that the software could predict and suggest artistic styles or modifications based on a user's past selections or based on the mood detected in the image.

Integration with Motion Graphics Beyond still images, there's a vast universe of motion graphics and videos waiting to be artistically transformed. Future versions of DeepArt Effects might seamlessly transmute videos, making them resemble hand-painted animations, much like the aesthetic of the film *Loving Vincent*.

Collaboration with Other AI-Driven Platforms DeepArt Effects could potentially collaborate with other AI-driven design tools. For instance, integrating AI-driven soundscaping tools could enable the creation of holistic multimedia art pieces, where the visuals and audio are both stylized in harmonious art styles.

Expansion into Wearable Tech With wearable technology such as smart glasses gaining popularity, there's potential for DeepArt Effects to provide real-time artistic filters for the world around us. Viewing the world through an impressionistic or cubist lens, quite literally, could be a fascinating application.

Crowdsourced Style Libraries While the platform might currently rely on renowned art styles, future trajectories could include a crowdsourced library where artists from around the world can upload their unique styles. Users could then access a global repository of styles, making the platform a nexus for international art collaboration.

Art Education and Training DeepArt Effects could become an educational tool, helping students understand the intricacies of different art styles by applying them to familiar images. This immersive learning experience could revolutionize art education, providing a hands-on approach to style appreciation.

More Ethical and Sustainable Art Creation Digital art platforms could potentially offer a more sustainable means of art creation, negating the need for physical materials, which can often be resource-intensive. Moreover, as copyright issues become more streamlined and clear-cut, the platform could ensure ethical art creation by flagging potential copyright infringements.

Enhanced Resolution and Detailing As computational capacities grow, future versions of DeepArt Effects could handle much higher resolution images, maintaining minute details even after the artistic transformation. This enhancement would be especially valuable for large-scale prints or digital displays.

Real-Time Feedback and Iteration Incorporating real-time feedback loops, where users can tweak settings and see instantaneous results, could make the art creation process more interactive and iterative. This feature would allow for nuanced adjustments, ensuring the final product is precisely as envisioned.

Expansion into Different Media The platform's core technology could be applied to different media, such as textiles or ceramics. Imagine wearing a dress or using crockery that mimics the artistry of renowned painters, all digitally printed based on DeepArt Effects transformations.

In summation, the intersection of art and AI, represented by DeepArt Effects, holds a kaleidoscope of possibilities. The future trajectories are not just about enhancing the platform's current capabilities but expanding its horizons to encompass various media, tools, and applications. As with any evolving technology, a balance between innovation and ethical considerations remains paramount. Still, there's no denying that the canvas for DeepArt Effects' future is vast and vividly imaginative.

26

Content Generation with AIVA

Table of Contents

Introduction to AIVA

AI, once the subject of science fiction, has rapidly found its place in a wide range of applications, from health care to transportation. However, one of the most intriguing applications of AI lies in the realm of creative arts. This brings us to AIVA (Artificial Intelligence Virtual Artist), an embodiment of the harmonious intersection between technology and music. AIVA is an innovative AI-driven music composer, a brainchild designed to craft compositions that mirror the intricacies of human-generated music, bridging the divide between machine precision and artistic expression.

At its core, AIVA is more than just a sophisticated algorithm; it represents a paradigm shift in the world of music composition. Trained on a vast array of classical compositions spanning centuries, from the ethereal notes of Bach to the profound symphonies of Beethoven, AIVA assimilates the essence of these musical masterpieces. Drawing from this extensive learning, it has the capability to generate original scores, tailored for a diverse range of media including films, video games, and commercials. The compositions are not mere replications or remixes of existing pieces, but are original creations, bearing testimony to the depth of its training and the potential of artificial intelligence in creative domains.

The inception of AIVA can be traced back to the burgeoning interest in employing deep learning techniques for various art forms. Deep learning, a subset of machine learning, involves training neural networks on vast datasets, enabling them to make decisions or predictions based on patterns. In the context of AIVA, this involved feeding it with countless hours of classical music, allowing it to recognize patterns, structures, and the nuances that define great compositions. The result? An AI composer capable of generating pieces of music that resonate with the emotional undertones often attributed solely to human creators.

A significant milestone in AIVA's journey came in 2016 when it became the first virtual composer to have its creations copyrighted. This was not just a win for AIVA but posed broader questions about the evolving landscape of creativity and ownership in the AI era. Can a machine be considered a legitimate composer? What does it mean for artists and the future of creative expression? While these questions linger, what's undeniable is AIVA's prowess and the doors it has opened for AI in artistic ventures.

However, AIVA isn't just a testament to technological advancement; it's a tool, an aid for human artists. For emerging filmmakers, game developers, or brands, the cost of original music compositions can be daunting. AIVA offers a solution—providing high-quality, original scores without the hefty price tag often associated with professional composers. It democratizes music composition, making it accessible to a wider audience and allowing for a richer tapestry of multimedia content.

In conclusion, AIVA stands at the forefront of the melding of technology and art. Its existence challenges preconceived notions of creativity, pushing boundaries and redefining what's possible. While the purists may argue that a machine can never replicate the soul and essence of human-made music, AIVA makes a compelling case for the vast potential that lies in the union of algorithms and artistry. As we stand on the cusp of this new era, one can't help but wonder what other horizons AI will expand in the world of creative arts.

How AIVA Utilizes Prompts

AIVA has revolutionized the intersection of AI and musical creativity. Its ability to generate intricate musical compositions is notable, but a pivotal element of its functionality lies in its interaction with user-defined prompts. Prompts serve as directive touchpoints, guiding AIVA's creative algorithms to align with specific moods, genres, or desired themes. Let's delve into the mechanics and implications of this dynamic.

Prompts as directional beacons: At its core, AIVA operates much like other deep learning models, utilizing vast datasets to derive patterns and generate outputs. However, without specific direction, the AI might produce generic or unspecific compositions. Prompts serve to channel AIVA's vast knowledge, ensuring the output aligns with a user's vision. For instance, a prompt such as "tranquil forest" might lead AIVA to generate a piece reminiscent of rustling leaves, bird calls, and a gentle breeze.

Layered interpretation: AIVA's strength doesn't merely lie in its ability to follow prompts but in how it interprets them. With its extensive training on diverse musical pieces, the AI can associate emotions, historical contexts, and intricate musical nuances with simple user prompts. Hence, "an eerie, suspenseful night" could draw influences from classical suspense themes, incorporating slow builds, haunting instrumentals, and abrupt note changes to create the desired mood.

Feedback loop integration: Beyond the initial prompt, AIVA allows for a degree of iterative feedback. Users can assess the initial output, refine their prompts, or offer feedback, which AIVA then processes for improved subsequent outputs. This feedback mechanism ensures that the AI's compositions evolve in alignment with user expectations, making AIVA more of a collaborative tool than a mere generator.

Bridging technical precision with emotional depth: Prompts are essential because they infuse AIVA's technical prowess with emotional depth. A generic AI-generated tune, though technically sound, might lack the emotional resonance that music demands. With prompts, users can guide AIVA toward capturing the desired emotional timbre, ensuring that the compositions are not just accurate but also emotionally evocative.

Flexibility across genres: AIVA's interaction with prompts is not constrained to a specific musical genre. Whether it's jazz, classical, rock, or electronic, AIVA utilizes prompts to traverse musical landscapes, drawing from its extensive training to generate pieces that resonate with the specified genre's characteristics. A prompt such as "futuristic techno" will yield vastly different results from "Renaissance court music."

Challenges of subjectivity: While prompts offer a way to guide AIVA, they come with inherent challenges. Music and its associated emotions are highly subjective. What might be joyful or melancholic to one user might have different connotations for another. This subjectivity can sometimes lead to a disparity between user expectations and AI outputs, emphasizing the importance of refining and clarifying prompts.

Prompts in the broader context: AIVA's usage of prompts can be viewed as a microcosm of the broader evolution of AI in creative domains. As AI ventures further into areas traditionally dominated by human creativity, prompts and similar directive mechanisms will play a crucial role. They serve as reminders that while AI can mimic patterns, the emotional depth, context, and nuance often require human input.

In conclusion, AIVA's interaction with prompts underscores a fundamental truth about the convergence of AI and art: machines can replicate patterns and structures, but the soul, emotion, and context often come from human intuition and direction. Prompts act as a bridge, channeling AIVA's computational capabilities into creations that resonate on a deeply emotional level. As AI continues to evolve, this harmonious blend of machine efficiency and human creativity, as exemplified by AIVA, promises to redefine the frontiers of artistic expression.

Examples of AIVA Applications

AIVA represents a leap forward in the world of AI-driven music creation, heralding an era where machines can craft complex and emotionally resonant pieces. Since its inception, AIVA has been applied in diverse domains, reflecting its versatility and the broad appeal of its generated content. Here are some salient examples of AIVA's applications:

Film scoring: One of the most prominent applications of AIVA is in the realm of film scoring. Filmmakers, especially those on tight budgets or schedules, have leveraged AIVA to generate background scores tailored to their movies' themes and moods. By providing prompts related to the film's storyline or desired ambiance, directors can obtain compositions that enhance the cinematic experience, without the traditional costs or timelines associated with custom compositions.

Video games: As the gaming industry continues to expand, the demand for unique and immersive soundtracks grows alongside. AIVA has been utilized to craft soundtracks for various video game genres, ranging from epic orchestral pieces for adventure games to eerie, atmospheric tunes for horror games. Its ability to iterate quickly also allows developers to test different musical styles and adapt based on gameplay feedback.

Advertisements: In the world of advertising, the right background score can significantly elevate a commercial's impact. AIVA has been used by advertising agencies to curate music that aligns with their campaign themes, allowing for a customized feel without resorting to generic stock music or expensive licensing.

Personal music creation: For aspiring musicians and hobbyists lacking in composition skills, AIVA has emerged as a valuable tool. Users can input themes or moods they wish to explore, and AIVA can generate pieces that they can then modify, expand upon, or integrate into larger works.

Music education: Educational institutions have incorporated AIVA into their curricula as a means to teach students about music theory, composition, and structure. By analyzing AIVA's creations, students gain insights into chord progressions, melody construction, and other foundational musical elements. Moreover, they can interact with the AI, tweaking inputs to see how compositions change, thus gaining a hands-on understanding of music creation.

Relaxation and meditation apps: The wellness industry, especially apps focused on relaxation, meditation, and mental well-being, have tapped into AIVA's capabilities. They utilize the AI to produce tranquil and calming tracks, aiding users in meditation, sleep, or relaxation exercises.

Event management: Event organizers, especially for occasions such as weddings, corporate events, or themed parties, have used AIVA to produce background scores tailored to the event's theme or the client's specifications. This customization adds a unique touch to the occasion, enhancing the overall ambiance.

Interactive art installations: In the realm of modern art, interactive installations that respond to viewers' actions or environmental factors are becoming popular. AIVA has been integrated into some of these installations to produce music in real time based on viewer interactions, making the art piece dynamic and ever-evolving.

Dance performances: Choreographers have utilized AIVA to craft original pieces around which they design their dance routines. This symbiotic relationship between AI-generated music and human movement showcases how technology and art can seamlessly intertwine.

Research and experimentation: Beyond commercial applications, AIVA has piqued the interest of researchers keen on understanding AI's potential and limitations in creative fields. Experiments where AIVA's compositions are compared with human-made pieces, or where its outputs are analyzed for emotional resonance and structural integrity, are not uncommon.

In conclusion, AIVA's reach extends far beyond mere music generation—it's reshaping how we perceive the act of creation itself. From the big screens of Hollywood to the intimate confines of a personal meditation session, AIVA's melodies echo, heralding a future where AI and human creativity coalesce in harmony.

Limitations of AIVA

AIVA, while a groundbreaking tool in the realm of AI-driven music generation, is not devoid of limitations and criticisms. As with many technological advancements, it stirs both awe and debate in equal measure. Here's a deep dive into some of the challenges and critiques associated with AIVA:

Lack of genuine emotion: The most prevalent criticism is that AIVA's compositions, while technically sound, often lack the genuine emotion and soul that human composers infuse into their works. Music is a deeply emotional art form, and many argue that AI-generated music, no matter how sophisticated, can't capture the depth and nuances of human sentiment.

Over-reliance on existing works: AIVA is trained on vast databases of preexisting music. While this enables it to generate compositions reminiscent of classical maestros or modern-day genres, it also means that it tends to produce music heavily influenced by what already exists. The question of true originality arises, with critics pointing out that AIVA's creations often echo familiar tunes.

Economic implications: There's a valid concern about AIVA replacing human composers in sectors such as film scoring, advertising, and video games. This could have significant economic implications for professionals in the music industry. Jobs and opportunities might dwindle as companies opt for cost-effective, AI-generated solutions.

Standardization and homogenization: With a tool such as AIVA becoming widely accessible, there's a risk that many creators might lean on it excessively, leading to a homogenization of musical content. This could stifle diversity and experimentation in music, making it harder for innovative and distinct sounds to emerge.

Ethical concerns of credit and ownership: As AIVA crafts compositions, the matter of crediting becomes complex. Who owns the rights to an AI-generated piece? Is it the developer of AIVA, the user who provided the prompt, or is it public domain? This uncharted territory raises legal and ethical concerns about ownership, copyright, and royalties.

Dependency and skill erosion: If emerging musicians and composers overly rely on AIVA for their creations, they might not hone their skills or tap into their intrinsic creativity. Overdependence on AI tools can potentially erode the very skills and talents the industry cherishes.

Lack of contextual understanding: While AIVA can be given prompts to guide its compositions, it doesn't truly understand the context. For instance, when scoring for a film, it doesn't grasp the storyline's depth, the characters' emotions, or the director's vision in the way a human composer would. This can lead to pieces that, while melodically apt, might miss nuanced cues critical to the scene.

Potential for misuse: With any tool as powerful as AIVA, there's always potential for misuse. There could be instances where individuals pass off AIVA's compositions as their own, leading to deceit and misrepresentation, especially if listeners aren't made aware of AI's role in the creation process.

Technical limitations: Like all AI models, AIVA isn't infallible. It can sometimes produce compositions that are discordant or lack coherence, especially when given vague or conflicting prompts. The outputs are only as good as the data it was trained on and the prompts it receives.

Cultural sensitivities: Music is deeply rooted in cultural contexts. A piece that's harmonious and pleasant in one culture might be jarring or inappropriate in another. AIVA, lacking cultural consciousness, might inadvertently produce music that's culturally insensitive or inappropriate.

In summary, while AIVA stands as a testament to the strides AI has made in the creative domain, it also brings forth a plethora of challenges and critiques. Balancing the potential benefits while addressing these concerns will be pivotal as we navigate the future of music in the AI era.

The Future of AIVA

AIVA represents a significant leap in the realm of AI-driven content creation, particularly in the domain of music composition. But as technology constantly evolves, so too will the capabilities and potential applications of AIVA. Let's explore the possible trajectories and implications for AIVA's future:

Integration with other media forms: As AIVA has already demonstrated prowess in music composition, the next logical step might be its integration with other forms of media. AIVA could be used in tandem with AI-driven video or animation tools, offering a cohesive audiovisual experience where music dynamically adjusts to visual elements in real time.

Personalized music creation: With user data becoming more nuanced and rich, AIVA might tailor compositions to individual tastes. Imagine a world where AIVA crafts a unique theme song for every individual based on their life events, moods, and preferences.

Live performance assistance: Beyond static compositions, AIVA might assist artists during live performances, dynamically adjusting background scores based on the performance's tone, the audience's response, or even current events.

Enhanced learning tools: AIVA could evolve into an educational tool, helping students learn music theory, composition, and instrumentation. By breaking down its compositions and explaining the rationale behind every note, it can offer a practical, interactive way for budding musicians to grasp complex musical concepts.

Expanded music database: As AIVA's training database grows to encompass a wider range of genres, cultures, and historical periods, its compositions will become more diverse. This could lead to a renaissance of forgotten or less-represented musical forms.

Collaboration with human artists: Instead of replacing human composers, AIVA might find its place as a collaborator. Artists could provide initial themes or ideas, which AIVA could expand on, offering a hybrid approach that combines human emotion with AI's computational prowess.

Ethical and regulatory developments: Given the complexities surrounding AI-generated content, the future might witness more stringent regulations about AI compositions' use and dissemination. Ethical guidelines might be established, emphasizing transparency, attribution, and copyright considerations.

Emotion detection and integration: Future iterations of AIVA could incorporate emotion-detection algorithms, allowing it to recognize and respond to human emotions in real time. This would enable the creation of music that resonates with the listener's current emotional state, offering therapeutic, entertainment, or immersive experiences.

Bridging cultural gaps: By training on a vast array of global musical genres, AIVA has the potential to create fusion compositions that blend diverse cultural elements. This could promote cross-cultural understanding and appreciation, presenting music as a universal language that transcends borders.

Addressing criticisms and limitations: AIVA's developers will undoubtedly refine the system in response to critiques. Enhancements in originality, reduction of over-reliance on existing works, and improvements in capturing nuanced emotional undertones might be focal areas of development.

Economic reconfigurations: As AI-driven music becomes more mainstream, the music industry's economic landscape might undergo shifts. New revenue models could emerge, focusing on AI-human collaborative projects, licensing of AI-generated music, or subscription models for continuous AI-driven music streams.

Interdisciplinary research: The nexus of music, AI, neuroscience, and psychology could be a future research frontier. By understanding how AI-generated music affects the human brain and psyche, we can refine AIVA's algorithms to produce compositions with specific cognitive or emotional impacts.

In essence, while AIVA has already made remarkable strides in AI-driven music composition, the journey has only just begun. The fusion of creativity and computation heralds a future where boundaries between human and machine blur, opening avenues previously deemed unimaginable. Embracing this potential while navigating inherent challenges will determine AIVA's trajectory in the ever-evolving music landscape.

Examples of AI Music Prompts (Source: https://contentatscale.ai/ai-music-prompt/)

Here are a few examples of AI music prompts for different musical genres:

Pop Music

"Create a catchy and upbeat chorus with a memorable hook."

"Generate a verse with a modern pop sound and relatable lyrics."

"Provide a danceable drum pattern for a pop party anthem."

"Design a synth riff that's both energetic and infectious."

"Generate a bridge that builds anticipation and leads to an explosive final chorus."

Classical Music

"Generate a melancholic and emotive violin melody in a minor key."

"Provide a regal and majestic brass section for a grand orchestral piece."

"Create a delicate and serene piano introduction for a sonata."

"Design a contrapuntal melody for a chamber music ensemble."

"Generate a hauntingly beautiful choir arrangement for a choral composition."

Hip-Hop/Rap

"Create a head-nodding beat with a deep bassline for a rap freestyle."

"Generate a catchy hook with rhymes that flow seamlessly together."

"Provide a trap-style drum pattern with hi-hats that add rhythm and groove."

"Design a verse with confident and clever wordplay."

"Generate a chorus that incorporates auto-tuned vocals for a melodic rap track."

Rock Music

"Create a powerful guitar riff that sets the tone for a rock anthem."

"Generate a hard-hitting drum pattern with driving rhythms for a rock ballad."

"Provide a dynamic and intense build-up for a guitar solo."

"Design a gritty and raw vocal melody for a rock song with attitude."

"Generate a bridge that transitions smoothly from a soft verse to a loud chorus."

Electronic/EDM

"Create an infectious and energetic synth melody for a club banger."

"Generate a drop that unleashes a burst of energy with pounding bass and synths."

"Provide a build-up that elevates the anticipation before a euphoric chorus."

"Design a glitchy and experimental intro for an electronic track."

"Generate a vocal chop sequence that adds a unique twist to the composition."

27

Audio Synthesis with WaveNet

Table of Contents

Understanding WaveNet

In the dynamic realm of audio synthesis, WaveNet emerged as a revolutionary paradigm. Developed by DeepMind, WaveNet is not just another iteration in the long lineage of audio synthesis methodologies; it signifies a fundamental shift in how machines understand

and generate sound. Born from the ambition to create more natural-sounding synthetic speech, WaveNet's design departs from traditional methods, gravitating toward a more intricate understanding of raw audio waveforms.

At its core, WaveNet is a deep generative model, specifically crafted for the generation of raw audio waveforms. The traditional methods of audio synthesis often relied on concatenative or parametric approaches, where prerecorded sound bites are stitched together or mathematical models try to approximate speech. These methods, while effective to a degree, often produced audio that sounded robotic or lacked the intricate nuances of genuine human speech. WaveNet, on the other hand, does not rely on prerecorded snippets or high-level approximations. Instead, it takes on the daunting task of generating sound one sample at a time, resulting in highly detailed audio outputs.

What sets WaveNet apart is its architectural foundation. It's constructed as a deep convolutional neural network (CNN), a structure more commonly associated with visual tasks in the realm of machine learning. However, in WaveNet, this architecture is optimized for the sequential and temporal nature of audio. A standout feature in its design is the use of dilated convolutions. As the depth of the network increases, the dilations double, allowing the model to have an exponentially increasing receptive field. This is crucial for audio synthesis, especially for human speech, where understanding the context (the sound bites before and after a point) can drastically change the meaning and sound of a word or phrase. Such an extensive receptive field means that WaveNet can grasp longer sequences of audio, capturing the subtleties and intricacies that make human speech sound natural.

Furthermore, the model's ability to be conditioned on external data augments its versatility. For instance, WaveNet can be conditioned on textual inputs, enabling text-to-speech applications. Similarly, conditioning it on different speaker identities makes it possible for the system to generate speech mimicking specific voices. This capability does not limit WaveNet to just speech; when conditioned on appropriate musical data, the model showcases potential in generating pieces of music, manifesting its wide-ranging applicability.

The training process for WaveNet is an exercise in patience and computational prowess. Given its deep architecture and the sheer granularity of generating audio sample by sample, the model demands substantial computational resources. However, once trained, the results are profoundly rewarding. The synthetic speech produced by WaveNet surpasses the quality of preceding methods, bridging the uncanny valley that often plagues synthetic audio, making it sound almost indistinguishably close to genuine human speech.

The rise of WaveNet heralded a new era in audio synthesis. Companies such as Google swiftly integrated WaveNet into their text-to-speech services, appreciating the unprecedented quality it brought to synthetic speech. Yet, like all innovations, WaveNet is not without challenges. The computational intensity of the model, especially in its early iterations, made real-time synthesis a challenge. Moreover, while the quality of the audio is exceptional, there can still be occasional artifacts or irregularities.

In summation, WaveNet represents a monumental stride in the journey of audio synthesis. By opting to understand and generate raw audio waveforms through a deep learning lens, it offers a glimpse into the potential of machine learning to recreate and possibly redefine the boundaries of sound. As technology continues to evolve, models such as WaveNet not only serve as benchmarks but also as inspirations, guiding the quest for even more authentic and diverse synthetic audio experiences.

The Significance of Prompts in WaveNet

In the world of machine learning and generative models, prompts act as the guiding lights, steering the generation process toward desired outcomes. For WaveNet, a deep generative model designed to produce raw audio waveforms, the concept of prompts or conditioning plays a pivotal role. It's through this conditioning that WaveNet can generate diverse sounds ranging from distinct human voices to varied musical tones. Let's delve deeper into the significance of prompts in WaveNet.

Contextual Conditioning

At its core, WaveNet is trained to predict the next audio sample based on previous samples. But when additional contextual information (prompts) is provided, the model becomes conditioned to generate audio in a specific context. For instance, providing a textual prompt can guide WaveNet to produce a corresponding speech waveform. This is the backbone of advanced text-to-speech applications using WaveNet.

Voice Modulation

One of the profound applications of prompts in WaveNet is voice modulation. By conditioning the model on speaker identities or voice profiles, WaveNet can emulate specific voices. This could range from mimicking the tonal qualities of male or female voices to replicating the unique voice of a specific individual. The conditioning data serves as a "voice fingerprint," guiding WaveNet's generation process.

Musical Generation

Beyond speech, WaveNet has shown capabilities in generating music. Prompts in this context can be particular instruments, musical genres, or even specific moods. By feeding the model with such prompts, it can be guided to produce a saxophone solo or a melancholic piano piece, demonstrating the versatility of WaveNet beyond mere speech synthesis.

Emotional Coloring

Speech isn't just about words; it's also about conveying emotions. Through specific prompts, WaveNet can be conditioned to produce speech with particular emotional undertones, be it joy, sadness, excitement, or anger. This becomes especially significant for applications such as virtual assistants or chatbots, where a more emotionally nuanced response can drastically enhance user experience.

Accent and Linguistic Nuances

Language and speech are deeply regional. The same words can be pronounced differently across regions, and this is where prompts become crucial. By conditioning WaveNet on specific regional or linguistic data, it can be guided to produce speech in a particular accent or dialect, enhancing its applicability across diverse linguistic landscapes.

Enhanced Training Efficiency

While WaveNet is designed to understand patterns in raw audio waveforms, introducing specific conditioning or prompts can make the training process more efficient. Instead of letting the model wander aimlessly in the vast ocean of possible audio sequences, prompts can act as anchors, ensuring that the model's exploration is more directed and purposeful.

Bridging Multimodal Domains

The idea of prompts in WaveNet opens the door to intriguing multimodal applications. Imagine conditioning the audio generation based on visual prompts, where WaveNet produces soundscapes or background scores for specific visual scenes. Such cross-modal conditioning can usher in new frontiers in content creation.

Limiting Overfitting

In machine learning, overfitting occurs when a model becomes too attuned to its training data, diminishing its performance on unseen data. Introducing varied prompts during training can act as a form of data augmentation, ensuring that WaveNet has a broader understanding and doesn't over-specialize.

However, while prompts play an integral role in guiding WaveNet's capabilities, it's crucial to approach them with care. Over-reliance on specific conditioning data can inadvertently narrow down WaveNet's generative capabilities. Moreover, the choice of prompts and conditioning data becomes critical from an ethical perspective, especially when replicating specific voices or producing content that can be perceived as original human creations.

In conclusion, prompts or conditioning in WaveNet are not just supplementary tools; they are foundational elements that unlock the model's versatility. They allow WaveNet to traverse the wide spectrum of audio, from specific voices to varied musical genres, ensuring that its synthetic audio output is not just accurate but also richly diverse and contextually relevant.

Showcasing WaveNet in Real-World Scenarios

Since its inception, WaveNet has revolutionized the field of audio synthesis, giving rise to numerous practical applications that touch various aspects of our daily lives. From virtual assistants to entertainment, its impact is profound. Here's an exploration of how WaveNet has manifested in real-world scenarios.

Virtual Personal Assistants

WaveNet's capacity to generate humanlike speech has elevated the experience of interacting with virtual assistants. Google Assistant, for instance, integrates WaveNet technology, allowing for smoother, more natural conversational interactions. Instead of the robotic tones typical of earlier voice assistants, WaveNet offers responses that sound nearly indistinguishable from a human, enhancing user engagement.

Audiobooks and Text-to-Speech Services

The traditional text-to-speech systems often lacked the emotional nuances of a human narrator, making long listening sessions somewhat monotonous. WaveNet changes this by introducing prosody and intonation variations, making audiobook listening a more immersive experience. Publishers and platforms can now convert text to speech without sacrificing the emotional depth that human narrators bring.

Entertainment and Gaming

The gaming industry demands diverse voiceovers for characters, often necessitating a vast cast of voice actors. With WaveNet, developers can generate a plethora of distinct voices, bringing characters to life with varying tones, accents, and emotions. This not only saves on production costs but also allows for on-the-fly voice generation, particularly useful in expansive open-world games.

Language Learning Apps

Pronunciation and intonation are crucial in language learning. WaveNet-powered applications provide learners with authentic, clear pronunciations, catering to multiple accents and dialects. This real-world sound mimicry accelerates comprehension and ensures learners can understand and be understood in genuine conversational settings.

Accessibility Tools

For the visually impaired, text-to-speech systems are vital. WaveNet's realistic voice outputs enhance the auditory experience for users relying on screen readers, making digital content more accessible. Furthermore, its potential in generating soundscapes can aid in developing tools that provide auditory cues about environments, helping users navigate unfamiliar spaces.

Music Generation

While WaveNet is primarily designed for speech, its deep learning foundations allow it to experiment with music. Artists and producers can harness WaveNet to create unique musical instruments or generate background vocals, pushing the boundaries of sound design.

Telecommunications

In an era of global connectivity, clear communication is paramount. WaveNet can be used in telecommunication systems to improve voice clarity, especially in scenarios with poor signal quality. Moreover, it can be integrated into voice response systems, providing callers with more natural-sounding interactions, reducing the "digital frustration" often associated with automated systems.

Film and Animation

Dubbing is a significant challenge in the film industry, especially when trying to maintain the emotional integrity of the original performance. WaveNet can be trained on actors' voices, enabling the generation of dialogue in multiple languages while preserving the original tone and emotion, ensuring a consistent viewing experience across languages.

Health Care

In scenarios where patients have lost their ability to speak due to medical conditions, WaveNet can be a beacon of hope. By training on previous voice samples, it's possible to recreate a patient's voice, allowing them to communicate through devices that generate speech mimicking their original tone, granting them a semblance of their former voice.

Training and Simulation

For professions requiring extensive communication training, such as customer support or sales, WaveNet can simulate diverse customer interactions, providing trainees with a variety of scenarios. This dynamic training ensures better preparedness for real-world situations.

While the applications of WaveNet are vast and varied, it's essential to approach its usage responsibly. The ability to mimic human voices brings about ethical challenges, especially concerning consent and potential misuse in generating misleading content. However, with appropriate safeguards, WaveNet promises to reshape the auditory landscape, bridging the chasm between digital interactions and genuine human conversations.

Limitations and Room for Improvement

WaveNet has undeniably pushed the boundaries of audio synthesis with its deep learning–based approach, delivering unparalleled voice quality and realism. However, like any pioneering technology, WaveNet is not without its limitations. Understanding these challenges can provide insights into the future evolution of audio synthesis. Here's an exploration of WaveNet's current constraints and areas awaiting enhancements.

Computational Costs

One of WaveNet's primary criticisms, especially during its early stages, was the extensive computational resources required for real-time voice generation. Though improvements have been made, the training process for the model, especially on vast datasets, remains resource-intensive, limiting its accessibility for smaller organizations or individual developers.

Latency Issues

For applications demanding real-time response, such as interactive voice assistants, any delay, however slight, can diminish the user experience. The autoregressive nature of WaveNet, where each sample is generated based on previous ones, can introduce latency. Optimizing this process remains a priority.

Data Dependency

WaveNet's proficiency is closely tied to the quality and diversity of the training data. In languages or dialects with limited available data, the model might underperform, leading to less realistic outputs. This dependency emphasizes the need for comprehensive and diverse datasets for training.

Ethical Implications

With WaveNet's capability to mimic human voices convincingly, there arises the potential for misuse, such as creating forged voice recordings. Addressing these concerns demands not just technological solutions but also regulatory oversight to prevent malicious applications.

Emotional Nuance

While WaveNet can generate humanlike speech, capturing the intricate emotional nuances of human conversation remains challenging. Ensuring the voice output aligns with the emotional intent of the text can make interactions more genuine and relatable.

Voice Individuality

Even with training on varied voices, there's a risk of synthesized voices sounding somewhat homogenized. Preserving the unique qualities of individual voices, especially when scaling to cater to millions of users, is an area where more refinement is needed.

Integration Challenges

Incorporating WaveNet into existing systems, especially those not originally designed for deep learning models, can be challenging. Seamless integration into various platforms and devices without compromising performance is crucial for widespread adoption.

Cost Implications

Leveraging WaveNet, especially for large-scale applications, can be expensive. From data acquisition to computational costs, and from model fine-tuning to deployment, organizations might find the expenses mounting. More cost-effective solutions can democratize access.

Interdisciplinary Collaboration

Audio synthesis, especially in applications such as health care, requires expertise beyond deep learning. For instance, creating a voice for a patient who lost their speech requires knowledge of vocal physiology, psychology, and more. Enhancing interdisciplinary collaboration can lead to more holistic solutions.

Environmental Impact

Training deep learning models on large scales has a carbon footprint. As the demand for voice synthesis grows, it's essential to consider sustainable and environmentally friendly computational practices to minimize this impact.

Model Interpretability

Deep learning models, including WaveNet, often suffer from being black boxes, meaning their decision-making processes are not transparent. Improved model interpretability can enhance trust and enable developers to fine-tune outputs more effectively.

Despite these challenges, WaveNet's contribution to the realm of audio synthesis is undeniable. By understanding its limitations and continuously working toward improvements, developers and researchers can harness its potential responsibly and innovatively. The journey of WaveNet, like any transformative technology, is one of iteration and evolution, with each step bringing us closer to blurring the lines between human and machine-generated audio.

The Next Steps for WaveNet

WaveNet, since its inception, has undoubtedly marked a revolutionary leap in audio synthesis, transforming voice generation with remarkable quality and realism. However, as technology progresses, the anticipation surrounding its future grows. Here's a glance into the prospective developments, enhancements, and broader horizons for WaveNet in the realm of audio synthesis.

Enhanced Real-Time Capabilities

Real-time voice synthesis is pivotal for various applications, including gaming, interactive voice assistants, and telecommunication. By refining the model's architecture and optimizing its algorithms, WaveNet can achieve faster generation speeds, thereby enhancing its real-time responsiveness.

Broader Linguistic Coverage

The world is linguistically diverse, and there are many languages and dialects yet to be explored by WaveNet. Expanding its linguistic repertoire will make the technology accessible and relevant to a broader global audience.

Emotional Depth and Nuance

Human communication is deeply intertwined with emotion. The next iterations of WaveNet might focus on not just mimicking human voice but also capturing and replicating the emotional nuances, offering a more genuine and emotionally resonant voice output.

Adaptive Learning

Future versions of WaveNet could employ adaptive learning, wherein the model fine-tunes itself based on user feedback. Such a self-improving model can consistently enhance its outputs, ensuring the synthesized voice remains top-tier over time.

Personal Voice Models

With advancements in data privacy and protection, individuals could potentially train personalized voice models using WaveNet. This could be revolutionary, especially for those who might lose their voice due to medical conditions, allowing them to communicate using a synthesized voice that mirrors their original one.

Reduced Environmental Impact

As discussions around the environmental impact of AI intensify, we can anticipate future iterations of WaveNet to be more energy efficient. Adopting green AI practices and sustainable computing will be vital steps forward.

Expanded Applications

Beyond voice assistants and music generation, WaveNet could explore new territories. Think of applications in film industries for generating background noises, health care for voice rehabilitation, or even in education for creating personalized learning assistants.

Holistic Multimodal Integrations

WaveNet's prowess in audio synthesis can be combined with advancements in other fields, such as video synthesis or virtual reality. This will lead to the creation of comprehensive multisensory experiences, where visuals and sounds are both AI-generated but indistinguishable from reality.

Ethical and Regulatory Frameworks

As WaveNet's voice synthesis becomes even more realistic, there will be a pressing need for robust ethical guidelines and regulations. Ensuring responsible use, preventing misuse such as deepfake audios, and establishing authenticity will be paramount.

Open-Source and Democratization

While proprietary advancements are crucial, the open-source community has always played a vital role in AI's evolution. By making more tools and versions of WaveNet available for public experimentation, we can witness a broader range of applications and innovations.

Intuitive User Interfaces

For non-tech-savvy individuals, using advanced models such as WaveNet can be daunting. The development of intuitive user interfaces and platforms will simplify the process, allowing more people to harness its capabilities without deep technical knowledge.

Collaborations and Partnerships

Interdisciplinary collaboration can open new doors for WaveNet. Partnerships with experts in fields such as psychology, linguistics, and entertainment can lead to innovative applications and ensure the technology evolves holistically.

The journey ahead for WaveNet in the domain of audio synthesis is filled with promise and potential. By addressing its current limitations and continuously pushing the boundaries, WaveNet is poised to redefine our auditory experiences. As AI continues its rapid evolution, the line between human-generated and AI-generated audio will blur, and WaveNet will undoubtedly be at the forefront of this transformative change.

28

Image Classification with ImageNet

Table of Contents

Getting to Know ImageNet

The emergence and evolution of ImageNet have forever transformed the landscape of computer vision and machine learning. At its core, ImageNet is a vast and meticulously curated dataset that consists of over 14 million annotated images, spanning 20,000 diverse categories. However, the significance of ImageNet extends beyond its sheer size or the diversity of its content. To truly grasp its impact, one must understand the convergence of ambition, innovation, and timing that marked its inception and the subsequent revolutions it inspired.

In the late 2000s, machine learning researchers grappled with the lack of extensive labeled datasets to train increasingly sophisticated models. It was during this period that Dr. Fei-Fei Li, then at Princeton and later at Stanford, envisioned a project that would tackle this challenge head-on. Motivated by the human brains rapid image recognition capabilities, her goal was to mimic such prowess computationally. Recognizing that large-scale data could be a game changer, the ImageNet project was initiated.

The crafting of ImageNet was no small feat. Collaborating with teams across the globe, Li utilized the WordNet hierarchy to define categories, or "synsets." These synsets were then populated with images sourced from the Internet, each of which was meticulously labeled by human annotators. This emphasis on accurate annotations ensured that the dataset was not just large but reliable.

But the real seismic shift came in 2012, a defining moment in the annals of deep learning. A convolutional neural network (CNN) named AlexNet, developed by Alex Krizhevsky, Ilya Sutskever, and Geoffrey Hinton, was trained on ImageNet. Their model's performance in the ImageNet Large Scale Visual Recognition Challenge (ILSVRC) shattered records, besting the previous year's results by over 10% in top-5 error rate. This was a watershed moment—the prowess of deep learning in handling image data was unequivocally showcased to the world.

After 2012, the ILSVRC became a hotbed of innovation. Year after year, models participating in this challenge, armed with the ImageNet dataset, pushed the boundaries of what was possible. Architectures such as VGG, GoogLeNet, and ResNet didn't just excel in this competition, but they laid foundational paradigms for a plethora of applications, from medical imaging to autonomous vehicles.

The ImageNet legacy extends beyond its dataset or the annual challenges. It's emblematic of a broader shift in machine learning—the movement from handcrafted feature engineering to letting neural networks learn features directly from data. This data-driven approach has since become a cornerstone of modern AI, influencing domains far removed from image classification.

Yet, the ImageNet journey is not without criticisms. Concerns have arisen over potential biases embedded within the dataset, mirroring broader conversations about fairness, accountability, and transparency in AI. There's also the consideration of overfitting; models might become so attuned to ImageNet that their real-world applicability becomes restricted.

In retrospect, ImageNet's influence is undeniable. While it stands as a testament to collaborative ambition and the power of large-scale, high-quality data, it also serves as a beacon, guiding discussions about the ethical implications and future trajectories of AI research. Looking ahead, as the field continues to evolve, ImageNet's foundational role will likely be revered, studied, and built upon by generations of AI practitioners and enthusiasts.

How Prompts Play a Role in ImageNet

ImageNet, with its colossal repository of labeled images, stands as a testament to the power of data in refining machine learning algorithms. However, the data alone isn't the sole factor behind ImageNet's success. The manner in which these images are labeled, categorized, and queried plays a critical role in the dataset's effectiveness. This brings us to the concept of "prompts"—structured inputs or cues used to guide a machine learning model's predictions or behaviors.

Defining Categories with WordNet

ImageNet's categorization is inherently a form of prompting. It leverages the WordNet hierarchy, a lexical database, to define categories known as "synsets." Each synset corresponds to a set of synonymous words or phrases, providing nuanced semantic context. When training models, these semantically rich categories offer a deeper understanding of the image content than mere generic labels would.

Training and Validation Phases

During the model training phase, images from ImageNet are used as input prompts, where a model is provided with an image and it has to predict the corresponding label. During validation, models are again "prompted" with new images, gauging their ability to generalize and predict accurate labels based on prior learning.

Large Scale Visual Recognition Challenge (ILSVRC)

ImageNet's annual competition, ILSVRC, also relies heavily on prompts. Participants are provided with specific tasks, such as object detection or scene parsing, and their models are evaluated based on their accuracy in responding to these prompts using the ImageNet dataset. The challenge prompts guide the direction of research, with teams focusing on achieving higher accuracy and efficiency in those specific areas.

Augmenting Data

ImageNet's success in training models also owes a lot to data augmentation techniques. These techniques, such as random cropping, rotation, or color jittering, modify the input images, effectively creating new "prompts" from existing data. By training on these augmented images, models become more robust and better at generalizing to diverse real-world scenarios.

Error Analysis and Model Refinement

After training, models might not always classify images correctly. Researchers use misclassified images as prompts in error analysis. By understanding where a model falters, they can refine the model's architecture, adjust hyperparameters, or even reframe the training prompts to address the shortcomings.

Bias and Fairness Assessments

Recent discussions around AI ethics have shed light on potential biases embedded in datasets such as ImageNet. By using specific images as prompts, researchers can evaluate a model's predictions to uncover and address any inherent biases. Such biases could range from racial and gender biases to subtle cultural nuances that the model might misinterpret.

Adversarial Testing

Adversarial images serve as specially crafted prompts designed to deceive machine learning models. When models trained on ImageNet are exposed to these prompts, they might misclassify them, even if the changes to the original image seem imperceptible to humans. These adversarial prompts play a pivotal role in testing the robustness of models and prompting further refinement.

In essence, prompts in the context of ImageNet serve as structured inputs or tasks that guide, test, and refine the behavior of machine learning models. Whether they come in the form of image categories, competition tasks, augmented data, or adversarial inputs, prompts help harness the vast potential of the ImageNet dataset. They provide direction to the training process, set benchmarks for performance, highlight areas for improvement, and ensure that the models' capabilities align with real-world expectations and requirements.

Practical Applications and Examples

ImageNet, initially envisioned as a vast database for academic research, has reshaped the landscape of computer vision and artificial intelligence. The inception and success of ImageNet's classification challenges have led to a renaissance in deep learning, particularly in convolutional neural networks (CNNs). Here, we delve into the practical applications and examples birthed from the advances facilitated by ImageNet.

Automated Image Tagging and Organization

Example: Google Photos: Leveraging deep learning models, platforms such as Google Photos can automatically categorize images, making it easier for users to search for specific photos based on content, such as "dogs" or "beaches." The foundation laid by ImageNet has been crucial in perfecting this categorization.

Health Care and Medical Imaging

Example: Diagnosing skin conditions: Models trained on large datasets similar in spirit to ImageNet can assist dermatologists by suggesting potential diagnoses for skin lesions or conditions, improving early detection rates of diseases such as melanoma.

Retail and E-commerce

Example: Visual search engines: Companies such as Pinterest and Amazon have developed visual search tools. Users can upload an image, and the system, backed by image classification algorithms, identifies similar products or pins.

Augmented Reality (AR)

Example: IKEA Place App: This AR app allows users to visualize how furniture would look in their home environment. Image classification helps the app recognize and understand the context of the room, ensuring the virtual furniture appears natural.

Agriculture

Example: Pest and disease detection: Drones equipped with cameras capture images of crops. These images, when processed by models similar to those trained on ImageNet, can identify signs of pests or diseases, allowing for timely interventions.

Autonomous Vehicles

Example: Tesla's Autopilot: Image classification models process the continuous stream of visual data from the vehicle's cameras to identify pedestrians, other vehicles, traffic lights, signs, and more, facilitating real-time decision-making.

Smart Surveillance

Example: Crowd management: During large public events or gatherings, smart surveillance systems can monitor crowd densities, recognize suspicious behaviors, or even detect unattended bags, enhancing security measures.

Natural Conservation

Example: Wildlife monitoring: In protected wildlife reserves, automated camera traps categorize images of passing animals. This aids biologists in tracking species populations, migrations, and behaviors without human intervention.

Manufacturing and Quality Control

Example: Defect detection: Cameras on manufacturing lines snap pictures of products, and image classification algorithms instantly determine whether an item has any defects or inconsistencies, ensuring high-quality standards.

Fashion and Apparel

Example: Stitch Fix's Style Shuffle: Users are shown clothing items and accessories. Their likes and dislikes feed into an image classification system, refining recommendations over time based on visual preferences.

Space Exploration

Example: Identifying geological features on Mars: NASA uses image classification to assist in processing the massive amounts of visual data sent back by rovers, helping scientists identify geological features of interest.

Facial Recognition Systems

Example: iPhone's Face ID: While more complex than basic image classification, Apple's Face ID system has roots in the same technology.

It recognizes and verifies the user's face to unlock the device, leveraging deep learning models for accuracy.

Entertainment and Media

Example: Video game environment generation: Modern video games use image classification to render realistic environments, understanding and categorizing in-game elements to provide players with an immersive experience.

The advancements in image classification, propelled by datasets such as ImageNet, have permeated virtually every sector, streamlining operations, enhancing user experiences, and opening doors to innovations previously deemed science fiction. The examples listed, though diverse, represent just the tip of the iceberg in the vast sea of ImageNet's real-world applications.

Limitations of ImageNet

ImageNet, while revolutionary in driving forward the fields of machine learning and computer vision, is not without its limitations. Understanding these constraints is crucial for researchers and practitioners who wish to extract maximum value from this tool while avoiding pitfalls. Here's a deep dive into the limitations of ImageNet.

Representation Bias

ImageNet's images predominantly come from the web, reflecting biases present in online content. This might lead to models that are skewed toward certain demographics, cultures, or contexts. For instance, images of certain animals may predominantly show them in a zoo rather than in their natural habitats, influencing how models perceive these animals.

Granularity Issues

Some categories in ImageNet are very specific, while others are broad. For instance, there are multiple breeds of dogs but only one category for humans. This uneven granularity can lead to models that are exceptionally good at differentiating between specific subcategories but struggle with broader categories.

Lack of Context

Images in ImageNet are stand-alone, often without a larger context. In real-world scenarios, understanding context is crucial for correct image interpretation. For instance, a model might correctly identify a ball but might not understand the game context in which it is being used.

Overfitting Risks

Given its prominence, many deep learning models are extensively trained and fine-tuned on ImageNet. This might lead to models that perform exceptionally well on ImageNet-related tasks but struggle in other real-world, diverse situations—a phenomenon known as overfitting.

Annotation Errors

With millions of images, ImageNet inevitably contains labeling errors. These errors, if not accounted for, can propagate and affect the accuracy of models trained on this dataset.

Static Nature

The visual world is dynamic, but ImageNet is relatively static, reflecting the state of the web at the time of its last update. New objects, styles, or phenomena emerging after its last collection might not be represented.

Ethical and Privacy Concerns

Some images within ImageNet might have been sourced without the explicit consent of individuals present in them. This poses ethical concerns, especially when these images are used for commercial or research purposes.

Limited Diversity in Environments

ImageNet's database doesn't adequately capture the wide variety of environments and lighting conditions present in the real world. A model trained on ImageNet might recognize an object in standard lighting but could falter under challenging conditions such as dim lighting or fog.

Cultural Biases

Given that a large proportion of the Internet's content is Western-centric, ImageNet might have cultural biases. This can lead to models misclassifying or misinterpreting objects, attire, or symbols from non-Western cultures.

Overemphasis on Visual Cues

ImageNet focuses exclusively on visual data. In the real world, humans often rely on multiple senses or cues, such as auditory or tactile, to understand and classify their environment. Models trained on ImageNet lack this multisensory understanding.

Limited to 2D Images

While ImageNet provides a vast array of 2D images, it doesn't encapsulate 3D, moving, or interactive objects. This limitation can hamper the application of ImageNet-trained models in augmented reality, virtual reality, or video analysis tasks.

Not Suited for All Domains

While ImageNet covers a broad range of categories, it's not exhaustive. Domains requiring specialized understanding, such as medical imaging or satellite imagery, might find ImageNet lacking in depth or relevance.

Complexity and Computation Demands

Training models on a massive dataset such as ImageNet requires significant computational power, making it inaccessible for hobbyists or researchers with limited resources.

In conclusion, while ImageNet has undeniably been a game changer in the world of AI, recognizing its limitations ensures that it is used judiciously. These constraints also present opportunities for the development of more refined, specialized, or diverse datasets in the future.

Future Developments in ImageNet

Since its inception, ImageNet has played a foundational role in the evolution of machine learning, especially in computer vision tasks. But as technology advances and our understanding of machine learning matures, it's clear that ImageNet too will need to evolve. Below are anticipated directions and developments that might shape the future of ImageNet.

Enhanced Diversity and Inclusivity

Recognizing the biases and limitations inherent in ImageNet's current dataset, there will likely be concerted efforts to expand its database with a focus on diversity. This could include capturing images from varied cultural, geographical, and socioeconomic backgrounds to ensure a more balanced representation of the global environment.

Improved Annotation and Verification

As the demand for more precise and complex machine learning models grows, the annotations accompanying images in ImageNet will need to become more accurate. Leveraging crowd-sourced corrections or automated verification systems can help rectify existing annotation errors and ensure higher accuracy for new images.

Incorporating Motion and Time

The next frontier for ImageNet might be to venture beyond static images. Incorporating short video clips can bring the aspect of motion and time, allowing models to understand dynamic scenarios, leading to richer training datasets for video analysis or action recognition tasks.

Augmented Reality (AR) and Virtual Reality (VR) Integration

With AR and VR becoming mainstream, future versions of ImageNet might include datasets optimized for these domains, enabling models to be trained for immersive environments and interactive scenarios.

3D Object Representations

There's growing interest in understanding objects in three dimensions, not just from a flat perspective. ImageNet could expand to include 3D scans or models of objects, facilitating training of models for 3D recognition tasks.

Embedding Multisensory Data

Future iterations of ImageNet might not just be restricted to visual data. Integrating multisensory data—such as sound or touch associated with images—could pave the way for multimodal machine learning models that have a more holistic understanding of the environment.

Frequent Updates and Expansions

To ensure relevancy, ImageNet will need to update its database frequently, reflecting the evolving digital landscape. This could mean more frequent crawls of the web or collaborations with other databases.

Specialized Subsets

Recognizing that not all tasks require the vastness of ImageNet, we might see the emergence of specialized subsets. These curated collections can focus on niche areas, such as medical imagery, aerial photos, or specific habitats, providing targeted datasets for specialized applications.

Enhanced Privacy Measures

In a world increasingly conscious of privacy, future versions of ImageNet will likely have stringent privacy checks in place. This might involve better vetting of images, ensuring no private data is inadvertently included, and possibly using techniques such as differential privacy to anonymize data.

Interactive Learning Platforms

Beyond being a mere dataset, ImageNet could evolve into an interactive platform where machine learning models can be trained in real time, receive feedback, and adjust accordingly. Such platforms can dramatically reduce the time needed to train sophisticated models.

Integration with Advanced Machine Learning Techniques

As techniques such as transfer learning, few-shot learning, or zero-shot learning become popular, ImageNet might be restructured or augmented to support such methodologies more efficiently.

Environmental and Ethical Responsiveness

Future ImageNet developments could be driven by ethical considerations, ensuring data collection is sustainable, non-intrusive, and respectful of local customs, beliefs, and regulations.

Open Collaboration with the Global Community

ImageNet's future might see more open collaboration, with researchers worldwide contributing to its expansion, refinement, and annotation, ensuring it remains a tool by the community, for the community.

In conclusion, while ImageNet's past has been monumental in shaping the AI landscape, its future promises even greater advancements. The challenges it currently faces are but stepping stones to a more inclusive, accurate, and comprehensive tool that will continue to power the next generation of machine learning innovations.

29

Video Synthesis with VQ-VAE

Table of Contents

Introduction to VQ-VAE

Vector quantized variational autoencoder (VQ-VAE) represents a nexus point in deep learning's journey, elegantly merging the worlds of autoencoders and vector quantization to pioneer a unique direction for generative modeling. At a foundational level, VQ-VAE builds on the traditional variational autoencoder (VAE) structure, introducing a critical distinction instead of relying on the continuous latent variables that characterize VAEs; VQ-VAE navigates toward discrete latent spaces, spawning a slew of advantages, especially in generative tasks.

Dissecting VQ-VAE necessitates an understanding of its two primary components. First is the VAE architecture. Variational autoencoders are generative models characterized by an encoder that projects input data into a latent space and a decoder that reconstructs the data from this latent representation. The latent space in a typical VAE is continuous, which, while offering smooth interpolations, can pose challenges in terms of optimization and does not naturally align with certain applications, especially those demanding discrete representations.

Enter the second component, vector quantization. Herein lies the key innovation of VQ-VAE. Instead of projecting data into a continuous latent space directly, the encoder's outputs are quantized to the nearest vector in a predefined codebook. Each of these vectors represents a specific pattern or feature. When an input passes through the encoder, it's essentially matched to the closest patterns from the codebook, leading to a discrete representation. During the decoding phase, these quantized values are utilized to regenerate the data.

The beauty of VQ-VAE is manifold. The discrete nature of its latent representation facilitates applications such as speech synthesis and video generation where discrete data forms, such as words or frame sequences, are intrinsic. The quantization mechanism also aids in combating some challenges faced by traditional VAEs, notably the notorious "blurry" reconstructions that arise due to the averaging effect in continuous latent spaces.

However, VQ-VAE is not just a theoretical marvel; its inception was motivated by practical challenges. In the realm of generative modeling, the ability to capture high-level semantics often requires a compromise on detailed, low-level information. VQ-VAE, especially its advanced iteration, VQ-VAE-2, addresses this by leveraging a hierarchical structure, quantizing data at multiple resolutions. This multitiered approach ensures that the model can capture both the broad strokes and the finer nuances.

Yet, the journey of VQ-VAE isn't devoid of hurdles. The quantization step, while groundbreaking, can lead to issues such as codebook collapse, where certain codebook vectors remain unused, reducing the model's expressive power. Furthermore, training VQ-VAE demands careful consideration. The quantization process introduces a non-differentiable step in the model, which means that traditional backpropagation can't be applied directly. To circumvent this, a straight-through estimator is often employed to approximate gradients during training.

In the grand tapestry of generative modeling, VQ-VAE stands out as a testament to the power of hybrid approaches. By synergizing the foundational principles of autoencoders and vector quantization, VQ-VAE offers a refreshing perspective on data representation and synthesis. Its hierarchical architecture, combined with the discrete nature of its latent space, positions VQ-VAE as an influential player in the ongoing saga of deep learning, promising richer, more detailed, and coherent generative outputs.

The Impact of Prompts on VQ-VAE

Video synthesis using models such as VQ-VAE can be influenced significantly by prompts, which act as high-level instructions or initializations that guide the model's generative process. This chapter delves into how prompts play an essential role in shaping the outcome and nuances of video synthesis using VQ-VAE.

Directing Content Creation

The primary advantage of using prompts with VQ-VAE in video synthesis is the ability to control and guide the generation process. A well-crafted prompt can direct the VQ-VAE to generate content that aligns closely with specific themes, subjects, or scenarios. For instance, providing a prompt such as "sunset over a calm sea" will guide the model to prioritize scenes and visuals that match this description.

Enhancing Specificity

While VQ-VAE can generate content based on its training, without a prompt, the results might be generic. Prompts act as a specificity tool, ensuring the generated content aligns with a particular vision or requirement. For instance, a vague request might lead to a general cityscape, but with a prompt like "New York City in the 1980s," the generated video will likely showcase iconic landmarks and the style of that era.

Influencing Temporal Dynamics

Video synthesis isn't just about generating visuals; it's about creating a sequence that unfolds over time. Prompts can dictate not just the content but also the temporal dynamics of a video. A prompt such as "A bird taking off in slow motion" informs the VQ-VAE about both the subject (a bird) and the desired temporal dynamics (slow motion).

Aiding in Hierarchical Generation

Advanced versions of VQ-VAE, such as VQ-VAE-2, employ a hierarchical structure. Prompts can be instrumental in guiding the model at each level of the hierarchy. For instance, at a higher level, a prompt can determine the overarching theme, such as "Rainforest," and at subsequent levels, refine details such as the specific animals or weather conditions present.

Bridging the Abstraction Gap

VQ-VAE operates in a discrete latent space, making it prone to capturing abstract features and patterns. While this abstraction is one of its strengths, it can sometimes produce content that's too generalized. Prompts help bridge this abstraction gap, anchoring the generative process in concrete, user-defined specifications.

Reducing Ambiguity

In the absence of prompts, VQ-VAE's generative process might be susceptible to ambiguities. For instance, if tasked with generating a "celebration" video, the model could produce anything from a birthday party to a parade. However, a prompt such as "Children's birthday party with a clown" reduces ambiguity, providing a clearer context.

Overcoming Data Limitations

While VQ-VAE learns from its training data, it's improbable for the model to be exposed to every possible scenario. Prompts can help overcome data limitations by guiding the model to mix and match learned patterns in novel ways, even if the exact scenario hasn't been encountered during training.

Counteracting Mode Collapse

Generative models can sometimes suffer from "mode collapse," where they generate only a subset of possible outputs. By continuously varying and refining prompts, users can ensure a broader exploration of the model's latent space, counteracting tendencies toward mode collapse.

Facilitating User Interaction

Lastly, prompts pave the way for interactive video synthesis. Users can iteratively refine prompts based on interim outputs, effectively having a "dialogue" with the model. This iterative feedback loop allows for the fine-tuning of generated videos until the desired output is achieved.

In conclusion, while VQ-VAE offers a powerful framework for video synthesis, prompts act as the rudder, steering the generative process with precision. They play a crucial role in enhancing specificity, reducing ambiguity, influencing temporal dynamics, and much more. As video synthesis technology continues to evolve, the interplay between models such as VQ-VAE and prompts will undoubtedly remain a focal point, ensuring that AI-generated content remains aligned with human intentions and desires.

Real-World Use Cases and Examples

The advent of models such as VQ-VAE has significantly revolutionized the domain of video synthesis. These deep learning models offer high-quality video generation capabilities that have been applied in various sectors. Let's explore some real-world use cases and examples where VQ-VAE is making a mark.

Entertainment and Media Production

The entertainment industry is one of the primary beneficiaries of VQ-VAE's video synthesis capabilities. Filmmakers can employ the model to generate background scenes or simulate crowd sequences. This technology can also be utilized for post-production enhancements, filling in missing visual data, or for creating special effects that would be expensive or challenging to achieve through traditional means.

Example: A filmmaker wants to showcase a historical setting, such as ancient Rome. Using VQ-VAE, the crew can generate authentic-looking scenes, incorporating details from the era without needing elaborate sets or CGI techniques.

Video Games Development

The dynamic nature of video games requires diverse and vast visual content. VQ-VAE can be utilized to generate realistic environments, character animations, or event sequences based on a set of prompts.

Example: In an open-world game, the developer can use VQ-VAE to generate varied terrains such as forests, mountains, or deserts, ensuring that players always encounter unique landscapes as they explore.

Educational Simulations

Educational institutions and e-learning platforms can harness VQ-VAE to create lifelike simulations, offering students a more immersive learning experience. Whether it's a reenactment of a historical event or a biological process, these simulations make abstract concepts tangible.

Example: A medical training program can utilize VQ-VAE to simulate surgeries or other medical procedures, allowing students to observe and learn in a controlled, virtual environment.

Advertising and Marketing

For advertisers looking to create captivating visuals, VQ-VAE offers a way to generate high-quality promotional videos. Brands can employ the model to visualize product concepts, simulate user experiences, or even generate content tailored to specific audiences.

Example: A travel agency wishing to promote exotic destinations can use VQ-VAE to create enticing videos of places they offer packages for, giving potential travelers a sneak peek of their next holiday.

Research and Development

In sectors such as automobile or aerospace, VQ-VAE can be instrumental in visualizing new designs, prototypes, or simulating scenarios to test product resilience and efficiency.

Example: An automotive company can use VQ-VAE to visualize how a new car design would look in different settings, from urban streets to rugged terrains.

Virtual Reality (VR) and Augmented Reality (AR)

The immersive nature of VR and AR requires a constant stream of high-quality visual content. VQ-VAE can generate realistic environments, objects, and scenarios, enhancing the user's virtual experience.

Example: In a VR-based training program for firefighters, VQ-VAE can create various fire scenarios, from a kitchen fire to a forest blaze, allowing trainees to experience and respond to different challenges.

Security and Surveillance

For security applications, VQ-VAE can help in simulating potential threats or scenarios, enabling agencies to prepare and strategize. Additionally, it can be used in forensic reconstructions.

Example: Security agencies can use VQ-VAE to generate simulations of potential security breaches, studying them to develop better countermeasures.

Fashion and Retail

Fashion designers and retailers can employ VQ-VAE to visualize new designs, patterns, or see how clothing items look in different settings or on various models.

Example: A fashion brand can use VQ-VAE to generate videos showcasing their new collection in different seasons or settings, from a sunny beach to a snowy landscape.

Art and Digital Creations

Artists and digital creators can harness VQ-VAE to experiment with visual narratives, generate unique artworks, or even create dynamic installations.

Example: An artist can input a basic theme or emotion into VQ-VAE, allowing the model to generate a video artwork that interprets and portrays that theme.

In conclusion, the capabilities of VQ-VAE in video synthesis hold vast potential across numerous sectors. Its ability to generate high-quality, diverse, and tailored visual content makes it a valuable asset in any field that relies on dynamic visuals. As the technology matures, its applications are bound to expand, further integrating synthesized video content into our daily experiences.

Limitations and Areas for Improvement

VQ-VAE represent a leap forward in the domain of video synthesis, offering the capability to generate high-quality videos through deep learning mechanisms. However, like all technologies, VQ-VAE has its limitations and areas that beckon further refinement. Let's delve into some of these challenges

Computational Demand

Challenge: VQ-VAE models, particularly when synthesizing video, demand significant computational resources. High-resolution video synthesis requires powerful GPUs and substantial memory.

Improvement: Continued optimization of the model and leveraging more efficient architectures can mitigate computational challenges. Also, advancements in hardware and cloud computing solutions can alleviate this concern over time.

Training Data Quality and Quantity

Challenge: The performance of VQ-VAE largely hinges on the quality and volume of training data. Poorly curated or insufficient datasets can yield subpar or unrealistic video outputs.

Improvement: Expanding and refining training datasets, coupled with techniques such as data augmentation, can enhance the results. Collaborative efforts in the community to share and develop diverse datasets can be beneficial.

Temporal Consistency

Challenge: Maintaining consistent temporal features across video frames can be challenging. This might lead to flickering effects or incongruities in synthesized videos.

Improvement: Integrating temporal consistency regularization during the training process or adopting models specifically designed for sequence data, such as recurrent neural networks (RNNs), can help maintain continuity.

Handling Multimodal Outputs

Challenge: Real-world scenarios may have multiple plausible video outputs. VQ-VAE might struggle to capture all these modalities, leading to a bias toward certain outcomes.

Improvement: Multimodal training methods and diversifying training data can aid the model in capturing varied possible outputs. Exploring techniques to allow user-guided synthesis might also offer solutions.

Generalization Across Domains

Challenge: A VQ-VAE trained on a specific domain or type of video might not generalize well to other domains, limiting its versatility.

Improvement: Techniques such as transfer learning, where a pre-trained model is fine-tuned for a new domain, can aid in achieving better generalization. Additionally, hybrid models combining features of different architectures might yield better cross-domain results.

Ethical and Misuse Concerns

Challenge: The ability to synthesize realistic videos brings about concerns of deepfakes and misinformation. Malicious actors can misuse VQ-VAE to generate misleading content.

Improvement: Incorporating digital watermarks or metadata can help track synthesized content. Additionally, educating the public and developing deepfake detection tools can mitigate risks.

Lack of Interpretability

Challenge: Deep learning models, including VQ-VAE, often function as black boxes. This makes it challenging to interpret or understand why certain outputs are generated, which is crucial for applications such as medical video synthesis.

Improvement: Research in explainable AI (XAI) can offer insights into making VQ-VAE more interpretable. Introducing transparency mechanisms can help users understand and trust the synthesis process better.

Overfitting to Training Data

Challenge: VQ-VAE might overfit to the training data, making it reproduce the training videos too closely and lacking in creativity or novelty.

Improvement: Regularization techniques, diverse training datasets, and monitoring model performance on validation data can help counteract overfitting.

Synthesis Time

Challenge: Real-time or rapid video synthesis might not always be feasible due to the inherent complexities of processing video data.

Improvement: Model optimization, hardware acceleration, and leveraging edge computing can speed up the synthesis process, inching closer to real-time outputs.

In summary, while VQ-VAE heralds a new era in video synthesis, its journey is far from complete. By addressing its current limitations and continually refining its capabilities, the future of VQ-VAE appears promising, potentially reshaping how we create and perceive video content. As with all technologies, the blend of research, ethical considerations, and practical applications will define its trajectory.

Looking Forward: The Future of VQ-VAE

Enhanced Computational Efficiency

As computational power increases and algorithms become more optimized, expect faster and more efficient training and synthesis with VQ-VAE.

Integration of quantum computing or specialized AI chips can revolutionize processing capabilities.

Improved Temporal Consistency

Future iterations of VQ-VAE will likely place a greater emphasis on achieving smooth, flicker-free video synthesis.

Merging architectures such as recurrent neural networks (RNNs) with VQ-VAE might be explored to better handle sequential data.

Expansion to Various Domains

Beyond entertainment and media, VQ-VAE could be applied to domains such as medical imaging, surveillance, and scientific visualization.

Cross-domain training might become a standard, allowing models to generalize better across varied types of videos.

Real-Time Video Synthesis

With advancements in both hardware and software, near-real-time or real-time video synthesis could become a reality.

This would open doors for interactive applications, gaming, and live events.

Addressing Ethical Challenges

Given the potential misuse for deepfakes, future VQ-VAE models might integrate built-in watermarking or metadata tagging to identify synthesized content.

Collaboration between technologists, ethicists, and policymakers will intensify to establish standards and guidelines.

User-Guided Synthesis

Future VQ-VAE applications might allow users to guide the synthesis process, specifying desired outcomes or characteristics.

This would make the technology more interactive and tailored to individual needs.

Enhanced Multimodal Outputs

Instead of being restricted to single outcomes, VQ-VAE might evolve to produce multiple plausible video outputs for a given prompt.

Techniques to better capture and represent multimodal distributions in data will be at the forefront.

Merging with Augmented Reality (AR) and Virtual Reality (VR)

VQ-VAE could play a pivotal role in generating dynamic content for AR and VR environments.

As AR and VR technologies gain traction, the demand for high-quality synthesized content will grow.

Improved Interpretability

As the need for transparency in AI models increases, future VQ-VAE models might integrate mechanisms for better interpretability.

Techniques from the field of explainable AI (XAI) could be incorporated into VQ-VAE architectures.

Continuous Learning and Adaptation

Instead of static training, VQ-VAE models of the future might continuously learn and adapt to new data.

This would keep the synthesized videos current, relevant, and in tune with evolving trends.

Integration with Other Modalities

Future applications might not restrict VQ-VAE to video alone. Integration with audio, text, and other data modalities could produce comprehensive multimedia outputs.

Such multimodal synthesis would be especially relevant for creating holistic virtual environments or simulations.

Emphasis on Personalization

As data becomes more personalized, VQ-VAE might be used to generate videos tailored to individual user preferences or behaviors.

Personalized video content could revolutionize advertising, entertainment, and education sectors.

Collaborative Synthesis

Future tools might allow multiple users to collaborate in real time using VQ-VAE, guiding and influencing the synthesis process collectively.

This would promote a more communal and interactive approach to video creation.

Open-Source Advancements

The AI community's emphasis on open-source resources might lead to more publicly available advanced versions of VQ-VAE, spurring innovation.

Community-driven enhancements can accelerate the evolution of VQ-VAE.

Sustainability Considerations

Given the environmental concerns related to large-scale AI models, future VQ-VAE iterations might focus on sustainability, optimizing for lower energy consumption.

Efforts will be made to make the technology more eco-friendly.

In essence, the horizon for VQ-VAE in video synthesis is expansive and filled with possibilities. As the technology matures and integrates with other advancements, its potential to redefine how we perceive, create, and interact with video content becomes increasingly evident. The synthesis of today is but a glimpse of what's on the cusp of tomorrow.

30

Your Future in Prompt Engineering

Table of Contents

Where Can Prompt Engineering Take You?

Prompt engineering, an emerging subfield in the vast world of artificial intelligence, is opening up an array of opportunities for tech enthusiasts, linguists, and creative thinkers alike. It's a discipline that marries the art of language with the precision of technology, and it's gaining traction rapidly. So where exactly can prompt engineering take you in the ever-evolving tech landscape? Let's explore.

Bespoke AI solutions for industries: Every industry has its own set of challenges, jargons, and nuances. A health care AI tool may need a completely different set of prompts than an AI serving the entertainment sector. Skilled prompt engineers can tailor AI models to meet the unique demands of various sectors, making them invaluable assets for companies aiming to deploy AI across their operations.

AI training and development: Just as a teacher guides a student's learning path, prompt engineers guide AI models. They play an instrumental role in training sessions, refining models to produce desired outcomes, making them pivotal for AI companies and research institutions focusing on the next-gen models.

AI communications and PR: As AI continues to pervade our lives, there's an increasing need for transparent communication about how AI thinks and operates. Those adept at prompt engineering can serve as AI communicators, explaining the intricacies of AI outputs to the public, stakeholders, and regulators.

Freelance opportunities: With the rise of platforms such as OpenAI, there's a growing market for freelance prompt engineers. They can create custom solutions for businesses, assist in academic research, or even serve as consultants for firms looking to integrate AI into their operations.

Content creation: The entertainment and content sectors can benefit immensely from AI, especially for tasks such as scriptwriting, game design, and virtual world creation. Prompt engineers can work closely with creative teams, using AI to brainstorm ideas or fill in content gaps.

Ethical and policymaking roles: As with all tech advancements, AI brings ethical dilemmas. How do we ensure AI-generated content respects boundaries? What biases might an AI model inherit, and how do we correct them? Professionals who understand the mechanics of AI, including the role of prompts, are well-positioned to guide discussions on ethical AI use and contribute to policymaking.

Educational roles: As the field grows, so does the demand for educators. Universities, online platforms, and training centers will need experts to teach prompt engineering to the next generation of AI enthusiasts.

Research and innovation: The world of AI is still relatively uncharted. Those diving deep into prompt engineering today are pioneers, setting the stage for discoveries and innovations. Research roles, both in academia and industry, beckon those with a penchant for exploration.

Business analytics and strategy: Understanding how to extract precise information from AI models can be a game changer for business analytics. Prompt engineers can work with data scientists and business strategists to derive actionable insights from vast datasets.

Global opportunities: AI is a global phenomenon. Proficiency in prompt engineering can open doors to international careers, collaborations, and projects. With AI models being used everywhere, from Silicon Valley to tech hubs in Asia and Europe, the world is truly an oyster for skilled prompt engineers.

In conclusion, prompt engineering is not just a technical skill; it's an art, a science, and a passport to numerous exciting career paths. As AI continues to shape our future, those who master the nuances of prompts will find themselves at the forefront of this technological revolution. Whether you're a language lover, a tech geek, or someone simply curious about the future, there's a place for you in the world of prompt engineering.

Potential Career Paths and Opportunities

Prompt engineering is an intersection of linguistics, technology, and human psychology. Its emergence in the AI realm underscores the need for enhancing AI models to better understand and respond to human inputs. As the world increasingly integrates AI into daily operations, prompt engineering is forging new career paths and opening a world of opportunities. Here are potential directions one can pursue.

AI Model Trainer The foundational role in prompt engineering involves training AI models to understand and react to various prompts. This position requires a keen understanding of both the AI's architecture and the desired output to create effective prompts.

AI Consultant Companies worldwide are integrating AI into their systems. As an AI consultant specializing in prompts, you can guide firms on crafting prompts for specific outputs, ensuring the AI integrates seamlessly into their operations.

Research Scientist The domain of AI is ever evolving. Being a research scientist in prompt engineering means delving deep into understanding how prompts influence AI and discovering methods to make this interaction more intuitive and effective.

AI Educator With the increasing interest in AI, educational institutions and online platforms seek experts to teach the nuances of AI, including prompt engineering. Here, you can shape the next generation of AI enthusiasts.

Ethics Officer in AI Given the power of AI, ethical considerations are paramount. Specialists in prompt engineering can work to ensure that AI models function within ethical boundaries, especially when generating content or making decisions.

Content Creator with AI Assistance The media, advertising, and entertainment sectors can benefit from AI-powered content. Prompt engineers can craft specific narratives or content pieces by guiding AI models through meticulously designed prompts.

Business Analyst In the business domain, extracting precise insights from AI can be transformative. A prompt engineer can work alongside data scientists to craft prompts that derive actionable business intelligence from complex datasets.

AI Product Manager Companies developing AI-based products need experts who understand the intricacies of AI interactions. As a product manager, you'll be responsible for guiding the AI's development, ensuring its responses align with user needs, largely determined by effective prompting.

Freelance AI Expert Many small and medium businesses wish to integrate AI but may not have in-house expertise. Freelancers with skills in prompt engineering can offer tailored solutions, crafting prompts that align with specific business objectives.

AI Interface Designer As chatbots and virtual assistants become commonplace, there's a need for professionals who can design intuitive AI interfaces. This role involves understanding user needs and crafting prompts that guide users and AI toward meaningful interactions.

In addition to these career paths, the realm of prompt engineering offers numerous tangential opportunities. For instance, one could specialize in creating prompts for specific industries, such as health care or finance, ensuring AI models in these sectors function optimally. Similarly, with the rise of AI in gaming, there's scope for experts who can guide game narratives using prompts.

The field of prompt engineering is still in its nascent stages, and as with any emerging domain, early entrants have the advantage of shaping its trajectory. As companies, researchers, and the public at large grapple with making AI more effective and user-friendly, prompt engineers will be at the forefront, turning AI's potential into tangible outcomes.

In summary, the world of prompt engineering is rich with possibilities. Whether you're diving deep into research, guiding businesses, or teaching the next generation, the skills acquired in prompt engineering will position you at the cutting edge of the AI revolution.

Building a Portfolio in Prompt Engineering

As the significance of AI in our daily lives grows, prompt engineering emerges as a vital component in harnessing the power of AI models. For those aspiring to make their mark in this domain, a portfolio can be an indispensable tool to demonstrate expertise, creativity, and understanding. Here's how you can build a compelling portfolio in prompt engineering.

Start with the Basics Before delving deep, ensure your portfolio provides a brief introduction to prompt engineering. This can include a concise explanation of what it is, its significance in the world of AI, and its potential applications. This sets the context for anyone reviewing your portfolio, even if they are not deeply familiar with the subject.

Document Your Process Prompt engineering is as much about the process as it is about the result. Showcase your methodology—from identifying a problem or a need, conceptualizing the prompt, testing, iterating, to the final optimized prompt. Demonstrating this process gives a glimpse into your systematic approach and problem-solving abilities.

Diverse Use Cases A variety of examples will illustrate the breadth of your skills. You might include prompts for textual tasks, media generation, data extraction, or even conversational AI. Each example should outline the objective and the result, providing a clear before-and-after scenario.

Showcase Success Stories If you've had particularly successful prompts, perhaps those that led to substantial improvements in AI outputs, be sure to highlight them. These act as testimonials to your expertise.

Highlight Iterative Improvements In some cases, the first prompt might not deliver the desired outcome. Showcase examples where you made iterative improvements based on AI feedback, demonstrating adaptability and persistence.

Incorporate Feedback Mechanisms Engaging with your audience is crucial. Allow for feedback on your portfolio. It not only provides an opportunity for continuous learning but also demonstrates your openness to collaboration and improvement.

Stay Updated with Industry Trends The world of AI is rapidly evolving. Dedicate a section of your portfolio to recent advancements in prompt engineering or AI in general. Reflect on these advancements, and if possible, incorporate examples where you've applied new methodologies or adapted to new trends.

Engage in Real-World Projects Whether it's freelancing, collaborating with research groups, or working on personal projects, real-world applications give your portfolio authenticity. Discuss the challenges faced, the strategies employed, and the outcomes achieved in these projects.

Include Peer and Client Testimonials If you've collaborated with others or provided prompt engineering solutions to clients, gather and include their testimonials. A third-party perspective can validate your skills and give potential employers or partners confidence in your abilities.

Reflect on Ethical Considerations In today's AI landscape, ethics is paramount. Dedicate a section to discuss ethical considerations in prompt engineering, showcasing your awareness and proactive approach in ensuring responsible AI usage.

A well-curated portfolio in prompt engineering not only demonstrates your technical acumen but also your holistic approach toward problem solving, ethics, and continuous learning. Remember, the goal is not just to showcase what you've done but to provide a glimpse into how you think, how you approach challenges, and how you envision the future of AI interactions.

For those aspiring to dive into the world of prompt engineering, a robust portfolio can be the key to unlocking opportunities, fostering collaborations, and staying at the forefront of AI-driven innovations.

Continuing Your Education and Skills Development

Prompt engineering, positioned at the intersection of linguistics, AI, and user experience, is an evolving discipline. As AI models become increasingly integral in myriad applications, the relevance and complexity of prompt engineering also expand. This dynamic nature of the domain necessitates continual education and skills development. Here's how one can approach this continuous learning journey in prompt engineering.

Formal Education and Courses Many leading universities and institutions now offer courses on artificial intelligence, machine learning, and related fields. Some courses are specifically tailored to natural language processing (NLP) and may delve into the nuances of prompting. Enrolling in these programs can provide a solid foundation.

Online Platforms and MOOCs Platforms such as Coursera, Udemy, and edX offer numerous courses on AI and machine learning. Some instructors even provide modules dedicated to prompt crafting, enabling learners to understand the latest methodologies from experts in the field.

Workshops and Boot Camps Regularly participating in workshops or boot camps can offer hands-on experience. These intensive training sessions, often led by industry professionals, provide insights into real-world challenges and solutions in prompt engineering.

Stay Updated with Research The world of AI is research-intensive. Following AI research platforms such as arXiv or Google Scholar, and keeping tabs on publications from conferences such as NeurIPS or ACL, can keep you updated on the latest advancements and methodologies in the domain.

Engage in Communities Join AI and NLP communities, forums, or groups on platforms such as Reddit, Stack Exchange, or Discord. Engaging in discussions, asking questions, and sharing experiences with peers can be immensely enlightening.

Hands-On Projects Theoretical knowledge must be complemented by practical experience. Regularly working on projects, experimenting with different prompts, and iterating based on outcomes can significantly sharpen your skills.

Feedback Mechanism Actively seek feedback on your prompts from colleagues, mentors, or community members. Feedback offers a fresh perspective and highlights areas of improvement, ensuring you remain on a growth trajectory.

Ethical Training As AI's influence continues to permeate society, ethical considerations become paramount. Take courses or attend seminars that focus on the ethics of AI. Understand the implications of prompts, potential biases, and ensure your work adheres to ethical guidelines.

Soft Skills Development While technical acumen is crucial, soft skills, particularly communication, are equally important. Being able to articulate the rationale behind a prompt or explaining complex AI behaviors in simple terms can be invaluable, especially when working in interdisciplinary teams.

Stay Curious Lastly, cultivate a mindset of curiosity. As AI models evolve, the techniques and nuances of prompt engineering will also undergo changes. Keeping an open mind, being willing to unlearn and relearn, and approaching challenges with curiosity will ensure you remain adaptable and relevant.

The field of prompt engineering, given its pivotal role in harnessing the potential of sophisticated AI models, demands a commitment to lifelong learning. As AI becomes ever more complex and its applications more widespread, prompt engineers will play a crucial role in ensuring that human-AI interactions remain meaningful, effective, and aligned with intended goals. Investing in continuous education and skills development is not just a recommendation; it's a necessity for those aspiring to excel in this dynamic and impactful domain.

Joining the Global Community of Prompt Engineers

The world of AI is vast and ever evolving, with prompt engineering emerging as a pivotal discipline at the crossroads of technology, linguistics, and user-centric design. Given its nascent and rapidly evolving nature, becoming part of the global community of prompt engineers can offer numerous benefits, from knowledge exchange to collaborative opportunities. Here's a deeper dive into how one can engage and thrive within this burgeoning global community.

Why Join? Firstly, it's essential to recognize the inherent value of being part of a global network. The community can provide fresh perspectives, expose you to diverse methodologies, and open doors to collaborative projects that might have been otherwise inaccessible. Such an environment fosters learning, innovation, and the shared goal of refining AI-human interactions.

Online Forums and Groups Websites such as Stack Overflow, Reddit, or specialized AI forums frequently have dedicated sections or groups for NLP, AI interactions, and prompt engineering. These platforms become repositories of real-world challenges, solutions, and discussions that can be invaluable for both novices and veterans.

Professional Organizations Consider joining organizations such as the Association for Computational Linguistics (ACL) or other AI-focused bodies. They often offer resources, conferences, and workshops specifically tailored for subdisciplines, potentially including prompt engineering.

Conferences and Workshops Globally recognized AI conferences, such as NeurIPS, ICML, or ACL, occasionally feature workshops or tracks dedicated to human-AI interactions. Attending these can provide insights into the latest research and trends in prompt engineering and allow for networking with field experts.

Open-Source Contributions The AI community thrives on open-source development. By contributing to or initiating open-source projects related to prompting and AI interactions, you not only build a tangible portfolio but also engage with a global cohort of like-minded engineers and researchers.

Local Meetups While the community is global, numerous local groups or meetups often discuss AI, machine learning, and related fields. Engaging at a local level can provide more frequent interactions, hands-on workshops, and the chance to build a strong network within your region.

Collaborative Research Being part of the global community creates opportunities for collaborative research. Teaming up with international peers can combine diverse perspectives and skills, resulting in richer research outcomes and broader reach.

Webinars and Online Training Many industry experts and organizations regularly host webinars and online training sessions on various AI sub-domains. Participating in these can offer both knowledge and an avenue to interact with global experts and attendees.

Stay Updated Technology and methodologies in AI and prompt engineering are fast-evolving. Regularly checking community-curated resources, newsletters, or journals can help you stay abreast of the latest developments and best practices.

Share and Teach As you gain expertise, consider giving back to the community. Hosting webinars, writing articles, or even mentoring newcomers can help consolidate your knowledge and position you as a thought leader within the global community.

In conclusion, joining the global community of prompt engineers isn't just about enhancing one's skills but is also about being part of a collective that's shaping the future of human-AI interactions. This interconnected ecosystem offers opportunities for growth, innovation, and collaboration, ensuring that as the field of prompt engineering advances, its practitioners grow and evolve with it. The shared experiences, knowledge, and goals of this global community are invaluable assets for anyone keen on making a mark in the world of prompt engineering.

Advanced Techniques: Staying Ahead in Prompt Engineering

Prompt engineering, as a budding domain within AI and natural language processing, is dynamically evolving. With increasing reliance on AI systems for varied tasks, ensuring that these systems understand and act on our intentions is crucial. Therefore, staying updated with advanced techniques in prompt engineering can help harness the power of AI more effectively. Here's a comprehensive look into how one can stay ahead in this field.

Understanding Neural Architectures As language models become more sophisticated, a deep understanding of the underlying neural architectures, such as transformer models, becomes vital. This helps in crafting prompts that align well with the model's internal mechanics.

Active Learning and Feedback Loops Integrate active learning processes where the model refines its understanding based on iterative user feedback. This feedback can be used to improve prompt designs iteratively.

Experimentation With evolving models, there's no one-size-fits-all approach in prompt engineering. Regular experimentation, A/B testing of prompts, and monitoring outcomes can help discover the most effective prompting strategies.

Contextual and Dynamic Prompting Instead of static prompts, employ contextual and dynamic prompts that adapt based on user behavior, external data sources, or situational requirements. This ensures more relevant and intuitive AI interactions.

Multimodal Interactions As AI systems evolve, they're not limited to text. Exploring prompts in multimodal scenarios, where inputs can be images, audio, or even gestures, becomes crucial. This challenges prompt engineers to think beyond words and consider a wide array of human-machine interactions.

Ethical Considerations Advanced prompt engineering should also involve ethical considerations, ensuring that prompts don't inadvertently lead to biased, discriminatory, or harmful AI outputs. Regular audits and understanding of potential ethical pitfalls are necessary.

Cross-Domain Expertise Given the myriad applications of AI, prompt engineers can benefit from knowledge in application domains such as health care, finance, or entertainment. This allows for crafting prompts that are not just technically sound but also contextually relevant.

Regular Training and Upgradation AI and NLP fields are rapidly advancing. Engage in continuous learning through online courses, workshops, and conferences that focus on the latest developments in prompt engineering and related areas.

Collaboration and Community Engagement Joining forums, discussion groups, or professional organizations can provide insights into the challenges faced by others and the solutions they've found. Collaborative projects can also lead to innovative prompt strategies that wouldn't emerge in isolated work.

Automation and Tools As prompt engineering grows, so do tools and platforms that assist in the process. Familiarity with platforms that offer automated prompting solutions, analytics, and feedback mechanisms can streamline the prompt engineering process.

In conclusion, prompt engineering is at the intersection of linguistic creativity and technical acumen. As AI models become more intricate and their applications more widespread, the art and science of guiding these models via prompts will continue to gain prominence. By staying updated with advanced techniques, aspiring and experienced prompt engineers alike can ensure they remain at the forefront of this exciting and powerful domain. Not only does this promise a bright career trajectory, but it also holds the potential to shape the future of human-AI collaboration in meaningful and positive ways.

Acknowledgments

I WOULD LIKE to acknowledge the support of everyone involved in this book, which serves as a beginner's guide to a topic that is just at the early beginnings of its evolution. I foresee many new platforms emerging that will possibly be more expansive and far reaching than what we are seeing today.

Thanks to my friends, family, and peers for constantly keeping me thinking about what the future can be.

Thank you to the whole team at Wiley for making this project come to fruition.

Thanks to the following individuals for voting on the various cover options and helping lock down the one we finally chose:

Amy Eddy
Aqil Khan
Franklin Samuel
Ian Horne
Manan Mehta
Mauricio Henrique Beccer

Michael Frick
Pooja Arun
Prem Thiagarajan
Saad Khan
Simran F.

Ian Khan—The Futurist

Ian Khan "The Futurist" is a multifaceted leadership expert, storyteller, and creative mind. Founder of AIRI™, the AI Readiness Initiative, aimed at helping organizations measure their current state of readiness for an AI-driven world, Ian has also created the Future Readiness Score™, a KPI, to scientifically measure how well an organization is ready to grow through unanticipated change. His work has been featured on CNN, BBC, Bloomberg, Fast Company, and other global media.

Ian is an authority on emerging technologies including artificial intelligence, Web3, metaverse, virtual reality, technology enablement, and cybersecurity. He speaks about digital transformation, algorithms, robotics, and automation with ChatGPT and generative AI as well as decentralization through blockchain from a leadership perspective.

As a storyteller Ian is passionate about using different media to help educate his audiences and has produced several critically acclaimed technology-focused documentaries available on Prime Video, Emirates Airlines, Tubi, FlyDubai, and other leading streaming platforms. Ian is deeply vested in technology education and future readiness and believes in the positive transformative power of technology.

Ian is also a multiple times published author and has written *Metaverse for Dummies*, the most comprehensive reference book to help readers of all levels understand virtual reality and the metaverse. Ian stars in a new truth-seeking documentary series *The Futurist*, an investigative and storytelling, episodic show to help understand the impact of emerging technologies on our world.

An outdoors person, Ian loves nature and currently lives in Toronto with his wife and two kids.

Learn more about Ian at www.IanKhan.com.